D1372004

BIOCOMPUTING

BIOCOMPUTING

Phillip A. Laplante (Editor)

Nova Science Publishers, Inc.
New York

F.W. Olin College Library

Senior Editors: Susan Boriotti and Donna Dennis
Coordinating Editor: Tatiana Shohov
Office Manager: Annette Hellinger
Graphics: Wanda Serrano
Editorial Production: Maya Columbus, Vladimir Klestov, Matthew Kozlowski and Lorna
               Loperfido
Circulation: Ave Maria Gonzalez, Vera Popovic, Luis Aviles, Raymond Davis,
               Melissa Diaz and Jeannie Pappas
Marketing: Cathy DeGregory

Library of Congress Cataloging-in-Publication Data
Available Upon Request

ISBN: 1-59033-889-8

Copyright © 2003 by Nova Science Publishers, Inc.
               400 Oser Ave, Suite 1600
               Hauppauge, New York 11788-3619
               Tele. 631-231-7269   Fax 631-231-8175
               e-mail: Novascience@earthlink.net
               Web Site: http://www.novapublishers.com

All rights reserved. No part of this book may be reproduced, stored in a retrieval system or transmitted in any form or by any means: electronic, electrostatic, magnetic, tape, mechanical photocopying, recording or otherwise without permission rom the publishers.

The authors and publisher have taken care in preparation of this book, but make no expressed or implied warranty of any kind and assume no responsibility for any errors or omissions. No liability is assumed for incidental or consequential damages in connection with or arising out of information contained in this book.

This publication is designed to provide accurate and authoritative information with regard to the subject matter covered herein. It is sold with the clear understanding that the publisher is not engaged in rendering legal or any other professional services. If legal or any other expert assistance is required, the services of a competent person should be sought. FROM A DECLARATION OF PARTICIPANTS JOINTLY ADOPTED BY A COMMITTEE OF THE AMERICAN BAR ASSOCIATION AND A COMMITTEE OF PUBLISHERS.

Printed in the United States of America

CONTENTS

INTRODUCTION

Since the mid-1990's advances in DNA sequencing have enhanced our understanding of humanity and all living things. Driven by these advances, the closely related sciences of Bioinformatics and Biocomputing have become the ultimate interdisciplinary study areas, forever blurring the lines between engineering, biology and computer science and bringing together researchers who ordinarily wouldn't interact.

While Bioinformatics largely focuses on the analysis, prediction, imaging and sequencing of genes, the broader, interdisciplinary field of Biocomputing includes the study of biological models of computing using traditional materials, genomic modeling and visualization, biomaterials for non-traditional computer designs and computer architectures for those materials. In addition, Biocomputing uses the principles and tools of computer science to model or algorithmically specify complex biological information systems and computational systems with life-like capabilities.

Biocomputing has manifested numerous government multi-agency programs, including the Human Genome Project, the High Performance Computing & Communications (HPCC) initiative, the Human Brain Project, and other related programs such as the National Information Infrastructure and Digital Libraries initiatives, which have strong bio-related components.

It is therefore, gratifying to observe, that in this volume a wide range of biocomputing research is represented. For example, in "Application of Neural Networks to the Segmentation of Microscopy Images" Guan provides an investigation of intelligent image processing algorithms to segment chromosomes in three-dimensional (3D) microscopy images using a confocal light microscope. The advantage of this approach is that it allows biologists to observe live (or preserved) dividing cells in 3D. Then, he shows how a model-based neural network can be used to improve the quality of the images, before using a self-organizing tree map (SOTM) to perform segmentation. This 3D approach to segmenting individual chromosome features preserves the 3D orientations in relation to the surrounding cell volume. Examples are provided to demonstrate the satisfactory performance of the proposed algorithms in the 3D case.

In "Parallel Computation for Coefficients of Determination in the Context of Multivariate Gene-Expression Analysis," Suh et al present a parallel program for assessing the codetermination of gene transcriptional states from large-scale simultaneous gene expression measurements with CDNA microarrays. This program is based on the coefficient of determination, which has been proposed for the analysis of gene interaction via multivariate expression arrays and the construction of genetic regulatory network models. Parallel computing is a key factor in the application of the coefficient of determination to a large set of genes owing to the large number of expression-based functions that must be statistically designed and compared. The parallel program, *parallel analysis of* gene expression (PAGE), exploits the inherent parallelism exhibited in the proposed codetermination methodology. Finally, an application to a Markovian regulatory network is given.

Next, in "Virtual Tissue Engineering of Cardiac muscle: Computational aspects," Clayton and Holden provide a novel computational model of the heart. This high performance computing and visualization application requires reformulation of existing algorithmic technologies. This research has important implications in heart arrhythmia research, the design and pre-screening of anti-arrhythmics, and defibrillation technologies, which is discussed.

In the next paper, "Combining Particle Swarms and k-Nearest Neighbors for the Development of Quantitative Structure-Activity Relationships," Cedeño and Agrafiotis describe the application of a new optimization technique, particle swarms, to develop quantitative structure-activity relationship (QSAR) models based on k-nearest neighbor and kernel regression. Particle swarms is a population-based stochastic method that has been used successfully for feature selection in which each individual explores the feature space guided by its previous success and that of its neighbors. Success is measured by the predictivity of the resulting model as determined by k-nearest neighbor and kernel regression. The swarm flies through the feature space in search of the global minimum, guided by the regression error. In this paper, the technique is evaluated using well-known QSAR data sets and compared to other machine learning techniques.

In "Towards Minimal Addition-Subtraction Chains Using Genetic Algorithms," Nedjah and Mourelle use genetic algorithms to find optimal addition-subtraction chains to compute power T^E, where T varies but E is constant, which is an NP-hard problem. Their solution is better than others that use genetic algorithms.

In "Dynamic DNA Computing Model," Qiu and Lu, Abstract: a new DNA computing model is introduced to solve the 3-Coloring problem. New algorithms are presented as vehicles for demonstrating the advantages of the new model, which can be expanded to solve other NP-complete problems. Moreover, this approach has the advantage of dynamic updating, so an answer can be changed based on modifications to the initial condition. The new also model makes use of this large memory resource by generating a "lookup table" during the process of implementing the algorithms. Finally, the new model has the advantage of decoding all the answer strands in the final pool very quickly and efficiently. The advantage provided by this new model makes DNA computing both efficient and attractive in solving computationally intense problems.

Finally, in "A One Instruction Set Architecture for Genetic Algorithms" Gilreath and Laplante demonstrate how the genetic algorithm can be implemented using a unique one instruction computer architecture. This application of one instruction computing provides a unique and natural combination as a general-purpose means to optimizing problems.

Moreover, the simplified construction using one instruction elements, lends itself well to computing in alternate materials such as organic, optical, chemical, or quantum components and nano-materials. In fact, the one instruction methodology has a biologic parallel in the creation of a one instruction computer using living cells, which is demonstrated.

Dr. Phillip A. Laplante, PE
Associate Professor of Software Engineering
Penn State University
Great Valley School of Graduate Professional Studies
plaplante@psu.edu

In: Biocomputing
Editor: Phillip A. Laplante, pp. 1-15

ISBN 1-59033-889-8
2003 © Nova Science Publishers, Inc.

Chapter 1

APPLICATION OF NEURAL NETWORKS TO THE SEGMENTATION OF MICROSCOPY IMAGES

Ling Guan

Department of Electrical and Computer Engineering
Ryerson University, Toronto, Canada M5B 2K3
lguan@ee.ryerson.ca

ABSTRACT

An investigation of intelligent image processing algorithms to segment chromosomes in three-dimensional (3D) microscopy images taken by a confocal light microscope is presented. The use of this confocal light microscope allows biologists to observe live (or preserved) dividing cells in 3D. However, the top and bottom surfaces of these image features are indistinct, therefore requiring feature enhancement and segmentation of the chromosomes. In the proposed approach, a model-based neural network is first used to improve the quality of the images, and then the newly proposed self-organizing tree map (SOTM) is applied to perform segmentation. Segmentation algorithms are developed to work both on 2D dataset, based on a projection of the three-dimensional dataset, and on 3D dataset directly. The 3D approach to segmenting individual chromosome features preserves the 3D orientations in relation to the surrounding cell volume. The proposed algorithms perform very satisfactorily in the 3D case. Examples are provided to demonstrate the performance of the proposed algorithms.

Key-words – microscopy, neural networks, segmentation.

1. INTRODUCTION

The interpretation of biological images of dividing cells is of vital importance in biological research. One way to improve the quality of such images is to use the confocal light microscope [1,2].

The introduction of the confocal light microscope improves the quality of acquired images. Due to limitation of the optical system, which introduces out of focus blurring, the top and bottom surfaces of features may still be indistinct even with the use of the confocal light microscope. Furthermore, in live tissue confocal microscopy, the automatic identification of boundaries and structures are very difficult because most cells are essentially translucent. Further improvement of the quality of the microscopy images is more likely to come from improved image processing, as the design of the microscopes is becoming a mature technology.

There are several aims to investigate image processing techniques for extracting features of chromosomes from live tissue in three-dimensional (3D) image datasets. Some of the most important are: a) to isolate the chromosomes from the background; b) to extract the coiled chromosomes and any cross-links found in the plant cell.

Previous work for extracting features of chromosomes in 3D microscopy images has mostly involved the use of transform techniques. In particular, a Hilbert transform is used to process the 2D image slices, and then a high pass boost filter along the line of sight (the z direction in Fig.1, and selective opacity based on intensity are applied to reconstruct the 3D image. Unfortunately this method presents some problems on rotations of the 3D object, showing ill-defined chromosomes through the z direction.

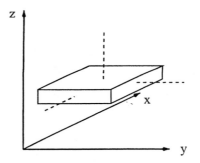

Figure 1: Object viewed from the 3 directions

This paper presents the application of neurocomputing methods to segment chromosomes in 3D microscopy images. The processing consists of two steps: a) regularize the distortions caused by imperfection in imaging and noise by a network of networks (NoN) [3,4]; b) segment the chromosomes using a newly developed self-organizing tree map (SOTM) [5,6]. The reasons for the use of the neural networks in this investigation are as follows:

- The nature of distortion in microscopy images is not well known. Therefore, a semi-blind deconvolution algorithm should be used for this regularization problem. It has been shown that the NoN is extremely suited for tasks like this [7,8]. What a NoN needs in restoration is the profile of the point spread function (PSF) and a very rough estimate of the PSF parameters. The NoN is able to adaptively search for the correct

PSF parameters and improve the quality of the images. NoN is also able to process images suffered from space variant distortion.

- Since there is not sufficient *a priori* knowledge of the number and types of regions to be separated, picture segmentation can be considered an unsupervised pixel classification problem, where the picture is divided into subsets by assigning the individual pixels to classes, while the classes themselves are determined by locating clusters in the feature space. The SOTM and other tree structured SOTM utilize powerful unsupervised learning methodology and are well suited for clustering and pixel classification [5,6,9].

The processing techniques are first developed to work on a 2D dataset, based on a projection of the 3D dataset, and then extended to 3D segmentation. The 3D algorithm is particularly useful in preserving 3D orientations in relation to the surrounding cell volume. Experiments show that the proposed method produces very satisfactory results.

2. THE NON FOR IMAGE REGULARIZATION

Sutton et. al. proposed the NoN neural network model which possesses a hierarchical cluster architecture [3]. It is a multi-level neural network consisting of nested clusters of neurons capable of hierarchical memory and learning tasks. In the network, each level of organization consists of interconnected arrangements of neural clusters. Individual neurons in the model form level zero of cluster organization. Local groupings among the neurons via certain types of connections, produce level one clusters. Other connections link the level one clusters to form level two, while the coalescence of level two clusters yields level three clusters, and so on.

Assuming linearity conditions, image regularization can be formulated as a clustering problem by the NoN. We utilize the popular quadratic error minimization model:

$$e(\mathbf{f}, \Lambda) = \frac{1}{2}\mathbf{f}^T(\mathbf{A} + \Lambda\mathbf{B})\mathbf{f} + \mathbf{c}^T\mathbf{f} + d \tag{1}$$

where $\mathbf{A}$ and $\mathbf{B}$ are matrices, $\mathbf{c}$ is a vector, d is a constant, $\mathbf{f}$ is the image in vector form, and Λ is a diagonal matrix representing adaptive regularization. $\mathbf{A}$, $\mathbf{B}$, $\mathbf{c}$ and d are functions of degradation and model statistics. Eq. (1) is an enhanced version of the popular formula used for restoration and statistical filtering [10].

Assuming there are S first-level clusters (the homogeneous areas in an image), we can link Eq. (1) to a three-level NoN model by first rearranging the image pixels in such a way that the pixels in a homogeneous image area (of arbitrary shape) are consecutively indexed in $\mathbf{f}$:

$$e(\mathbf{f}_*, \Lambda_*) = \frac{1}{2}\mathbf{f}_*^T(\mathbf{A}_* + \Lambda_*\mathbf{B}_*)\mathbf{f}_* + \mathbf{c}_*^T\mathbf{f}_* + d \tag{2}$$

where

$$\mathbf{f}_* = [\mathbf{f}_1^T \mathbf{f}_2^T ... \mathbf{f}_S^T]^T \tag{3}$$

is the rearranged image vector $\mathbf{f}$, with $\mathbf{f}_k$ being the vector representation of the pixels in the k th homogeneous area (first-level cluster), and

$$\mathbf{A}_* = \begin{bmatrix} \mathbf{A}_{11} & \mathbf{A}_{12} & \cdots & \mathbf{A}_{1S} \\ \mathbf{A}_{21} & \mathbf{A}_{22} & & \mathbf{A}_{2S} \\ \vdots & & \ddots & \vdots \\ \mathbf{A}_{S1} & \mathbf{A}_{S2} & \cdots & \mathbf{A}_{SS} \end{bmatrix}, \mathbf{B}_* = \begin{bmatrix} \mathbf{B}_{11} & \mathbf{B}_{12} & \cdots & \mathbf{B}_{1S} \\ \mathbf{B}_{21} & \mathbf{B}_{22} & & \mathbf{B}_{2S} \\ \vdots & & \ddots & \vdots \\ \mathbf{B}_{S1} & \mathbf{B}_{S2} & \cdots & \mathbf{B}_{SS} \end{bmatrix}, \mathbf{c}_* = \begin{bmatrix} \mathbf{c}_1 \\ \mathbf{c}_2 \\ \vdots \\ \mathbf{c}_S \end{bmatrix}, \tag{4}$$

and

$$\Lambda_* = \begin{bmatrix} \lambda_1 I & & & \phi \\ & \lambda_2 I & & \\ & & \ddots & \\ \phi & & & \lambda_S I \end{bmatrix} \tag{5}$$

are the corresponding rearranged $\mathbf{A}$, $\mathbf{B}$, $\mathbf{c}$ and Λ, respectively.

Then Eq. (2) can be rewritten as

$$e(\mathbf{f}, \Lambda) = \sum_{k=1}^{S} \frac{1}{2} \{ \mathbf{f}_k^T (\mathbf{A}_{kk} + \lambda_k \mathbf{B}_{kk}) \mathbf{f}_k + \mathbf{c}_k^T \mathbf{f}_k \}$$
$$+ \sum_{l=1}^{S-1} \sum_{l=k+1}^{S} \frac{1}{2} \{ \mathbf{f}_k^T (\mathbf{A}_{kl} + \lambda_k \mathbf{B}_{kl}) \mathbf{f}_l + \mathbf{f}_l^T (\mathbf{A}_{lk} + \lambda_l \mathbf{B}_{lk}) \mathbf{f}_k \} + d \tag{6}$$

where $\mathbf{A}_{kk} + \lambda_k \mathbf{B}_{kk}$ represents intracluster connections within cluster k, and $\mathbf{A}_{kl} + \lambda_k \mathbf{B}_{kl}$ and $\mathbf{A}_{lk} + \lambda_l \mathbf{B}_{lk}$ represent intercluster connections between cluster k and cluster l. Eq. (2) is the vector representation of the three-level NoN in [3]. The concept has been extensively studied and resulted in the development of adaptive image regularization theory and systems [11].

3. THE SOTM AS AN IMAGE SEGMENTOR

Vector quantization can be regarded as a mapping from a n-dimensional Euclidean space onto a finite set of prototypes. Algorithms for prototype generation and clustering attempt to organize unlabeled feature vectors into natural clusters in such a way that the entities within a cluster are more similar to each other than those in different clusters, and to represent them compactly with one or more prototypes. The K-means algorithm and the self-organizing map (SOM) are considered as two popular algorithms for such purposes [12]. However, each of the two has its own weakness. The K-means does not preserve topological relations in the input space. For the SOM, when the input vector distribution has a prominent maximum shape, the results of the best-match computations tend to be concentrated on a fraction of nodes in the map. This may cause statistical instability in the design of the segmentation algorithms [12].

Based on the idea of the self-organizing map, we recently proposed a new representation method called the self-organizing tree map (SOTM) in which the relationships between the output nodes are defined adaptively during learning [5]. The motivation for the SOTM is different from Kohonen's motivation of the original SOM – a nonparametric regression model, but it is an effective tool for pixel classification leading to segmentation and other image processing tasks. The clustering algorithm starts from an isolated node and coalesces the nearest patterns or groups according to a hierarchy control function from the root node to the leaf nodes to form the tree as shown in Fig. 2. The SOTM mapping projects an input pattern $x = (x_1...x_N) \in R^N$ onto a tree node.

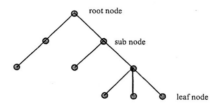

Figure 2: Structure of the SOTM three hierarchy

With every node j, a weight vector $w_j = [w_{1j},...w_{Nj}]^T \in R^N$ is associated. This method can accurately locate cluster centers, and preserve the topology of the input data. The SOTM provides a better and faster approximation of prominently structured density functions inherent in the input data. The SOTM method is summarized in the following algorithm:

[Step 1.]   Initialize the weight vector with a random value (We randomly take a training vector as the representative of the root node).

[Step 2.]   Get a new input vector, and compute the distances d_j between the input vector and all the nodes using

$$d_j = \sqrt{\sum_{i=1}^{N}(x_i(t) - w_{ij}(t))^2} \qquad (j = 1,...J) \qquad (7)$$

where J is the number of nodes.

[Step 3.]   Select the winning node j^* with minimum d_j.

$$d_{j^*}(x, w_j) = \min_j d_j(x, w_j) \qquad (8)$$

[Step 4.]   **If** $d_{j^*}(x, w_j) \leq H(t)$ where $H(t)$ is the hierarchy control function which decreases with time, and which controls the number of tree levels, **then** assign x to the j_{th} cluster, and update the weight vector w_j according to the following learning rule:

$$w_j(t+1) = w_j(t) + \alpha(t)[x(t) - w_j(t)]; \qquad (9)$$

where $\alpha(t)$ is the learning rate which decreases with time, and satisfies $0 < \alpha(t) < 1$.

Else form a new sub node starting with x.

[Step 5.]   Repeat by going back to Step 2.

Learning in the SOTM takes place in two phases: the locating phase and the convergence phase. The adaptation parameter $\alpha(t)$ controls the learning rate which decreases with time as weight vectors approach the cluster centers. It is given by an exponential function $\alpha(t) = e^{(-t/T_1)}$, where T_1 is a constant which determines the decreasing rate. During the locating phase, global topological adjustment of the weight vectors w_j takes place. The parameter $\alpha(t)$ is maintained relatively large during this phase. Initially, $\alpha(t)$ may be set at close to 1 and allowed to decrease with time. After the locating phase, a small $\alpha(t)$ for the convergence phase is needed for the fine adjustment of the map.

The hierarchy control function $H(t)$ controls the number of tree levels. It begins with a large value and decreases with time. It adaptively partitions the input vector space into smaller subspaces. The function $H(t)$ is defined as $H(t) = e^{(-t/T_2)}$ where T_2 is a constant which controls the rate of decrease. Both T_1 and T_2 are set at large values ($\approx 100,000$) in order to avoid missing the optimal (global) solution to convergence.

With the decreasing of the hierarchy control function $H(t)$, a subnode forms a new branch. The evolution process progresses recursively until it reaches the leaf node. The entire tree structure preserves topological relations from the root node to the leaf nodes.

The SOTM, in fact, is much better than the SOM at preserving the topological relations of the input dataset. This dynamic SOTM topology is best demonstrated in the following example. The learning of the tree map of Fig. 3 is driven by sample vectors uniformly distributed in the English letter "K". The tree mapping starts from the root node and gradually generates its sub nodes. The SOTM finds the cluster center just as K-means does.

Figure 3: English letter "K"

During vector quantization, an N-dimensional vector in Euclidean space is approximated by its closest representative among the finite set of the tree nodes (reference vectors). From visualizing the tree map evolution in Fig. 4a) to c), we can see that the optimal number of output nodes has been obtained through this node organization process. From this figure, it can also be seen that all the tree nodes are situated within the area of the distribution. The entire tree faithfully reflects the distribution of the input space. Using these nodes as code vectors, the vector quantization distortion can be kept to a minimum. For comparison, the SOM is also used to work on this example. Fig. 4d) shows that although the topology of the SOM exhibits the distribution of the structured input vectors, it also introduces false representations outside the distribution of the input space due to its regression nature.

This example shows that, unlike the SOM which attempts to cover the whole image through nonparametric regression, the SOTM provides a sparse, but accurate encoding of the prominent input patterns in the scene. This is very relevant to segmenting chromosomes in 3D microscopy images.

4. 3D TO 2D PROJECTION

In segmentation, a 2D scheme is first developed in order to minimize the computational complexity. We consider a projection of the 3D dataset onto one single slice. Due to the distortion introduced during the acquisition process, the relationship along the 3 axes is considered. By viewing the object from any of the three directions in Fig. 1, a mapping from three dimensions onto two dimensions is performed. As a result, two dimensions are preserved, but the dimension along the line of sight (LOS) is lost. The representation of the information in the lost dimension in the remaining dimensions determines the appearance of the object.

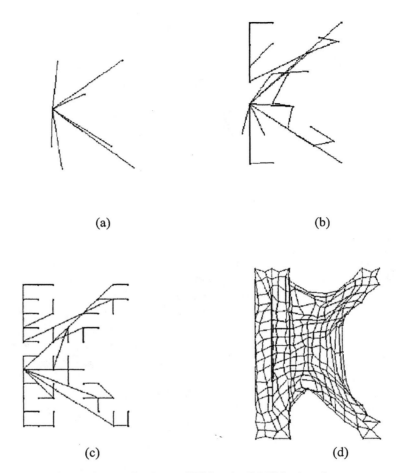

(a)   (b)

(c)   (d)

**Figure 4: learning letter "K" by the SOTM: a) and
b) the evolution the SOTM tree map; c) the final
representation of the SOTM; d) the representation of "K" by the SOM.**

Projection algorithms have been used in vision and image processing mainly for the purpose of simple but effective representations of the voluminous data encountered in processing, especially in 3D cases [13]. The project method used in this work is a simplified version of th maximum point projection algorithm we developed [14]. This method uses the number of voxel values along the line of an arbitrary direction to represent the projection onto the plane perpendicular to that line. It is noted that the simplified maximum point projection algorithm is more effective than those described in [13] due to the ill-conditioned nature of the microscopy images.

The projection method works as follows. Consider the projection of the object from three dimensions to two dimensions. Let (i, j, k) be a reference point on the observer plane. The vector normal to the observer plane is oriented horizontally by an angle θ and vertically by an angle ϕ. Let (i', j', k') be a point on the observer plane satisfying:

$$\cos\phi\cos\theta(i'-i)+\cos\phi\cos\theta(j'-j)+\sin\phi(k'-k)=0 \tag{10}$$

Let the set $LOS'(i',j',k') = \{x',y',z'\}$ be the set of all pixels along the direction of the normal vector to the observer plane at position (i',j',k'), where:

$$x' = R\cos\phi\cos\theta + i'$$
$$y' = R\cos\phi\sin\theta + j'$$
$$z' = R\sin\theta + k' \tag{11}$$
$$(0 \le R < \infty)$$

where R is the distance from the observer plane.

Consider a two dimensional coordinate system on the observer plane such that (i,j,k) is considered the origin. Let (α,β) be a point on the plane such that the mapping from the observer plane coordinates to the image space coordinates are:

$$(i',j',k') = \Gamma(\alpha,\beta) \tag{12}$$

Using the above mapping let $LOS(\alpha,\beta,\theta,\phi)$ be the line of sight from the observer plane at position (α,β) along the direction of (θ,ϕ). To extract the main body of the image $F(x,y,z)$, first the image is threshold

$$F_1(x,y,z) = \begin{cases} 0, & \text{if } F(x,y,z) \le \lambda; \\ 255, & \text{otherwise} \end{cases} \tag{13}$$

where λ is a pre-defined threshold value. Then a two-dimensional "density map" of the image is created whereby a projection onto the two-dimensional observer plane was made. The two-dimensional density map is generated as follows

$$D(\alpha,\beta) = \sum_x \sum_y \sum_z \left(\frac{F_1(x,y,z)}{255} \right) \tag{14}$$
$$\forall(x,y,z) \in LOS(\alpha,\beta,\theta,\phi)$$

Since the image has already been threshold to values of either 0 or 255, equation (14) states that the density map is created by counting the number of nonzero pixels along the line of sight from the observer plane. Despite of the complexity of (14), the implementation is very simple when the line of sight is along one of the x, y or z axes, just counting the number of nonzero voxels in $F_1(x,y,z)$. This density map D serves as the 3D to 2D projection to facilitate the segmentation of the microscopy images.

5. EXPERIMENTAL RESULTS

5.1. Feature Selection

In segmentation, a feature vector **x** consisting of three simple features: the pixel value, the mean, and the variance, and three more elaborated features: the median value within the window, the difference between the median and the current pixel value, and the connectivity count surrounding the current pixel. The mean value is taken as the average of the values in a 3x3 window and in a 3x3x3 window surrounding the current pixel itself, in the 2D and in the 3D cases, respectively. The variance value is given by the equation:

$$s^2 = \frac{\sum_{i=1}^{N} (x_i - \mu)^2}{N} \tag{15}$$

where N is the size of the window and μ is the local mean. As shown in the following, the utilization of this set of features actually achieved significant improvement over the method based on Hilbert transform.

5.2. Segmentation of Microscopy Images

Two datasets were used in the evaluation of the neural network algorithms. The image in Fig. 5a) is one of the 18 slices of the small dataset. The size of each slice is 256×256. It can be seen that the image is quite blurry, as are the rest of the slices in the dataset. The features in the chromosomes, as well as the chromosomes themselves, are quite indistinct.

As the first step, the NoN was used to regularize the image slices. Then the SOTM algorithm was applied to extract the chromosomes in both 2D and 3D cases. The image in Fig. 5b) is the result of the maximum point projection algorithm ran on the 18 slices dataset. It showed an enhancement of the chromosomes from the background. Even though the image seems to be less noisy, details and edges of the chromosomes are still not well defined. Segmentation is yet needed, but the histogram in Fig. 5c) does not give a good indication of how to separate the chromosomes from the background by conventional methods.

We therefore applied the SOTM to segment the image in Fig. 5b). The segmented result displayed in Fig. 5d) reveals much more information about the chromosomes, their position and their features, as compared with the original slices or with the projected image. The chromosomes are well defined from the background, i.e. edges are clearly visible, and the features of the chromosomes are properly emphasized.

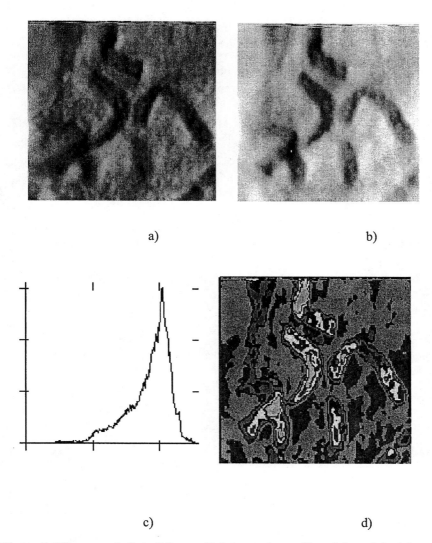

a)   b)

c)   d)

Figure 5: 2D segmentation of the small dataset: a) one slice of the original dataset composed by 18 slices, taken from a confocal light microscope; b) result of the projection algorithm on the 18slices dataset; c) histogram of b); d) 2D segmentation of the image in b).

There are reasons why this result is still not sufficient for analyzing the images. The most important is that by reducing the 3D dataset to two dimensions, we lose information on the 3D specimen object from which we want to extract features.

Fig. 6a) is a 3D view of the same dataset produced using a 2D Hilbert transform, a high pass boost in the z direction, and selective opacity based on intensity. In addition the out of focus planes were cropped. The image shows a good isolation of the general shape of the chromosomes. The twin strands of the chromosomes are visible, together with twisting. However, on rotating this 3D visualization, it becomes apparent that the chromosome objects are not well defined through z. In addition there is significant background noise. Imaging of a much thicker dataset in z would be problematic without better definition of the object boundaries in all the 3 dimensions.

Fig. 6b) show the 3D segmentation result obtained by the SOTM. The opacity has been selected based on the neural net mapping. The boundary between the object and background can therefore be made sharper and clearer. Large-scale features like chromosome twist and its 3D orientation still correspond to those seen in Fig. 6a) and in the original dataset, but the objects are now much more localized in the z axis, giving richer chromosome features which correspond to expected biological morphology.

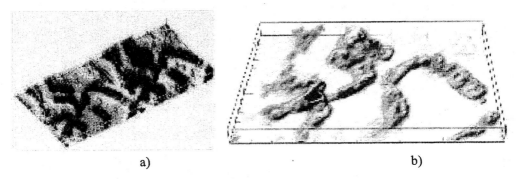

a)   b)

Figure 6: 3D segmentation and reconstruction of the small dataset: a) 3D view of the original dataset processed as described in the introduction; b) 3D view of the same dataset processed by the proposed method.

The image in Fig. 7a) is one of the 83 slices of the large dataset from a plant cell with groups of chromosomes in the center. It is hard to visualize how the chromosomes are distributed, and how many of them are there because they all clustered together in a big 'lump'. Hilbert transform based approach failed in this case. Apparently the 2D segmentation approach will also fail due to the overlap of the chromosomes. To make the problem more complicated, the image looks quite blurry. To process this dataset, the NoN was first applied to deblur the distortion. Although the nature of the distortion is not completely known, the profile of the PSF is approximately a Gaussian function

$$H(x,y) = K \exp \frac{x^2 + y^2}{\sigma^2} \tag{16}$$

where σ control the effective width of $H(x,y)$ and K is a scaling constant. Applying the NoN using the PSF given in (16), a clearer picture is obtained. By utilizing the 3D algorithms on the regularized image, the details of the surface of the chromosomes are successfully extracted and satisfactorily defined as shown in Fig. 7b).

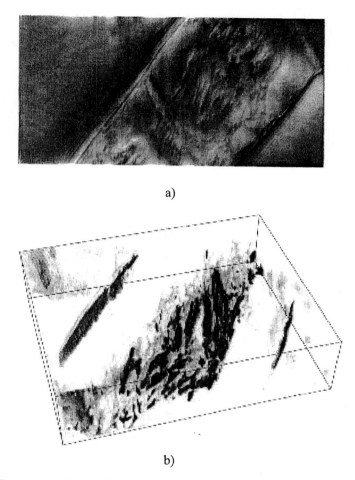

a)

b)

Figure 7: 3D segmentation and reconstruction of the large dataset: a) one slice of the large dataset composed of 83 slices, taken from a confocal light microscope; b) 3D view of the same dataset processed by the SOTM.

The figure reveals that there indeed exist multiple layers of chromosomes all twisted around one another. This could not be observed in the 2D segmentation result.

6. CONCLUSIONS

The difficulties of interpreting biological images have been greatly reduced by the use of the confocal light microscope. Even so, there are limits to what this microscope can do. Image processing is then needed to further improve the quality of the images. The techniques used in this work to improve image quality are model-based neural networks for regularization and self-organizing tree map for segmentation. The results produced in these experiments have been a great success, especially the 3D segmentation result. Two-dimensional segmentation gives a good feature extraction when there is only one layer of chromosomes in the 3D dataset because the chromosomes' features are very well clustered. Viewing the 2D segmentation result does not provide much information about the imaged specimens' 3D structure or even fail to provide any meaningful improvement, especially when multi-layers of chromosomes exist. The generalization of the SOTM to 3D segmentation is then proposed. This approach has preserved the 3D orientations in relation to the surrounding cell volume. Moreover, the structure of the features is better defined.

ACKNOWLEDGMENTS

The authors like to thank Dr. Stuart Perry for helping to develop the Maximum Point Projection Algorithm.

REFERENCES

[1]   C. J. Cogswell, K. G. Lakan, J. W. O'Byrne and M. R. Arnison, "High resolution, multiple optical mode confocal microscope: I. system design, image acquisition and 3D visualization", *SPIE Proc. Series*, vol. 2184, pp. 48-54, 1994.

[2]   A. E. Dixon and C. J. Cogswell "Confocal Microscopy with Transmitted Light", *Handbook of Biological Confocal Microscopy, 2nd Ed.*, Editor J.B. Pawley, pp. 479-487, 1995.

[3]   J.P. Sutton, J.S. Beis, and L.E.H. Trainor, "A hierarchical model of neocortical synaptic organization," *Mathl. Comput. Modeling*, vol. 11, pp. 346-350, 1988.

[4]   T. Caelli, L. Guan and W. Wen, "Modularity in neural networks," *Proc. of the IEEE*, vol. 87, no. 9, pp. 1497-1518, Sept., 1999.

[5]   H. Kong, A Self-Organizing Tree Map for Image Processing, *Ph.D. Dissertation*, University of Sydney, January, 1999.

[6]   J. Randall, L. Guan, W. Li and X. Zhang, "Hierarchical cluster model for perceptual image processing," to be presented at *IEEE Int. Conf. on Acoustics, Speech and Signal Processing*, Orlando, USA, May 2002

[7]   S.W. Perry and L. Guan, "Weight assignment of neural networks in adaptive image restoration," *IEEE Trans. on Neural Networks*, vol. 11, no. 1, pp. 156-170, January, 2000.

[8]   H.S. Wong and L. Guan "Fuzzy neural networks for image regularization," *IEEE Trans. on Neural Networks*, vol. 12. no. 3, pp. 516-531, May 2001.

[9]   S.K. Halgamuge, "Self-evolving neural networks for rule-based data processing", *IEEE Trans. on Signal Processing*, vol. 45, no. 11, pp. 2766-2773, Nov. 1997.

[10]   W.K. Pratt. *Digital Image Processing*. John Wiley & Sons, New York, 2nd ed, 1991.

[11]   S.W. Perry, H.S. Wong and L. Guan, *Adaptive Image Processing: A Computational Intelligence Perspective*, CRC Press, Boca Raton, 2002.

[12]   T. Kohonen, *Self-organizing Maps*, Springer-Verlag, Berlin. 1995.

[13]   D.H. Ballard and C.M. Brown, *Computer Vision*, Prentice Hall, Englewood Cliffs, NJ, 1982.

[14] S.W. Perry and L. Guan, "Segmentation and visualization of 3D underwater ultrasonic images," *Proc. APRS Image Segmentation Workshop*, pp. 69-74, edt Braun M et al, University of Technology Sydney, Australia, 1996.

In: Biocomputing
Editor: Phillip A. Laplante, pp. 17-28

ISBN 1-59033-889-8
2003 © Nova Science Publishers, Inc.

Chapter 2

PARALLEL COMPUTATION FOR COEFFICIENTS OF DETERMINATION IN THE CONTEXT OF MULTIVARIATE GENE-EXPRESSION ANALYSIS

Edward B. Suh,[1] Daniel E. Russ,[1] Edward R. Dougherty,[2] Seungchan Kim,[3] Michael L. Bittner,[3] Yidong Chen,[3] and Robert L. Martino[1]

[1]Division of Computational Bioscience, Center for Information Technology,
National Institutes of Health
Bethesda, Maryland 20892
[2]Department of Electrical Engineering
Texas A&M University
College Station, Texas 77843
[3]Cancer Genetic Branch, National Human Genome Research Institute,
National Institutes of Health
Bethesda, Maryland 20892
suhe@mail.nih.gov

ABSTRACT

This paper presents a parallel program for assessing the codetermination of gene transcriptional states from large-scale simultaneous gene expression measurements with cDNA microarrays. The parallel program is based on the coefficient of determination, which has been proposed for the analysis of gene interaction via multivariate expression arrays and the construction of genetic regulatory network models. Parallel computing is key in the application of the coefficient of determination to a large set of genes owing to the large number of expression-based functions that must be statistically designed and compared. The parallel program, *Parallel Analysis of Gene Expression* (PAGE), exploits the inherent parallelism exhibited in the proposed codetermination methodology. An application to a Markovian regulatory network is given

Key-Words: - parallel computing, data visualization, coefficient of determination, gene expression, cDNA microarray.

1. INTRODUCTION

An important goal of functional genomics is to develop methods for determining ways in which individual actions of genes are integrated in the cell. One way of gaining insight into a gene's role in cellular activity is to study its expression pattern in a variety of circumstances and contexts, as it responds to its environment and to the action of other genes. Recent methods facilitate large scale surveys of gene expression in which transcript levels can be determined for thousands of genes simultaneously. In particular, cDNA microarrays result from a complex biochemical-optical system incorporating robotic spotting and computer image formation and analysis [1,2]. Since transcription control is accomplished by a method which interprets a variety of inputs [3-5], we require analytical tools for expression profile data that can detect the types of multivariate influences on decision-making produced by complex genetic networks. Kim *et al.* have proposed using the coefficient of determination (CoD) for finding associations between genes [6-8]. The method assesses the codetermination of gene transcriptional states based on statistical evaluation of reliably informative subsets of data derived from large-scale gene-expression measurements with cDNA microarrays. The CoD measures the degree to which the transcriptional levels of a small gene set can be used to predict the transcriptional state of a target gene in excess of the predictive capability of the mean level of the target. In addition to purely transcriptional features, the method allows incorporation of other conditions as predictive elements, thereby broadening the information that can be evaluated in modeling biological regulation. The CoD has also been used for choosing gene predictor sets and determining their selection probabilities in the design of probabilistic Boolean networks [9], which form a special class of Markovian regulatory networks that generalize the classical Boolean genetic networks [10,11].

Data from cDNA microarrays are preprocessed before CoD calculation occurs. An algorithm calibrates the data internally to each microarray and statistically determines whether the data justify the conclusion that an expression is up- or down-regulated [12]. The complexity of expression data from a microarray is reduced by quantizing changes in expression level into ternary values: -1 (down-regulated), $+1$ (up-regulated), or 0 (invariant). The number of gene from the microarray is reduced from several thousand to several hundred by requiring at least c changes in the quantized gene expression data over the samples. For the data set used in this paper, $c = 4$.

Application of the statistical framework to a large set of genes requires a prohibitive amount of computer time on a classical single-CPU computing machine. To meet the computational requirement, we have developed a parallel implementation of the codetermination method, named the *Parallel Analysis of Gene Expression* (PAGE) program.

2. CODETERMINATION ALGORITHM

The genomic regulation patterns seen in microarray data can be viewed as a biological system S with a k-input expression level vector $X = \{X_1, X_2, ..., X_k\}$. The target Y is the output expression value of system S. A nonlinear predictor L is constructed to estimate Y, and it can be either a perceptron or a ternary logic filter. The results from the optimal predictor Y_{pred} are the closest approximation to Y. The logic of L represents an operational model of our understanding of the biological system [6-8].

The theoretically optimal predictor has minimum error across the population. It is unknown and must be statistically estimated; that is, it must be designed from a sample by some training (estimation) method. How well a designed predictor estimates the optimal predictor depends on the training procedure and the sample size n. The error ε_n of a designed predictor must exceed the error ε_{opt} of the optimal predictor. For a large number of microarrays, ε_n approximates ε_{opt}, but for the small numbers typically used in practice, ε_n may substantially exceed ε_{opt} [13].

The data problem can be mitigated if, instead of estimating the best predictor, we estimate the best predictor from a constrained set of predictors. The theoretical error of a best-constrained predictor exceeds that of the best predictor; however, the best-constrained predictor can be designed more precisely from the data. Hence, the error of a designed estimate of an optimal constrained predictor is often less than the error of a designed estimate of the optimal unconstrained predictor. This paper focuses mainly on the full-logic (unconstrained) predictor.

The CoD of the optimal predictor θ_{opt} is the relative decrease in error owing to the presence of the observed variables:

$$\theta_{opt} = \frac{\varepsilon_b - \varepsilon_{opt}}{\varepsilon_b} \tag{1}$$

where ε_b is the error for the best predictor in the absence of observations. Since the error, ε_{opt}, of the optimal predictor cannot exceed ε_b, $0 \le \theta_{opt} \le 1$ [6-8]. If optimization is relative to *mean-square-error* (MSE), which is the expectation $E[|Y_{pred} - Y|^2]$, then the best predictor based on the predictor variables $X_1, X_2, ..., X_k$ is the conditional expectation of Y given $X_1, X_2, ..., X_k$. The best predictor of Y in the absence of observations is its mean, μ_Y, and the corresponding error is the variance of Y. Predicting Y by its mean might yield small or large MSE because the variance of Y is small or large, respectively. Thus, there is normalization by ε_b in Eqn. 1 to measure the effect of the observations. In our setting, the conditional expectation needs to be quantized to ternary values, and ε_b is the error from predicting Y by applying the ternary threshold of the mean of Y. The CoD measures normalized predictive capability in a probabilistic, not causal, sense. A high CoD may indicate causality of the predictors on the target or vice versa. It may also indicate diffuse control along network pathways, or co-regulation of both predictors and target by another mechanism. It is why the CoD is being called the *codetermination* in biological applications.

For designed predictors, ε_b is replaced by $\varepsilon_{b,n}$ the error resulting from using the ternary threshold of the sample mean of Y to predict Y, and ε_{opt} is replaced by ε_n to give the estimate θ_n of the CoD:

$$\theta_n = \frac{\varepsilon_{b,n} - \varepsilon_n}{\varepsilon_{b,n}} \qquad (2)$$

We use cross-validation to estimate θ_n. The data is split into training and test data, a predictor is designed from the training data, and estimates of $\varepsilon_{b,n}$ and ε_n are obtained. An estimate of θ_n is found by putting the error estimates into Eqn. 2. The procedure is repeated a number of times and a final estimate, $\hat{\theta}_n$, is obtained by averaging. $\hat{\theta}_n$ is a conservative estimate of θ_{opt}. We could also use resubstitution to estimate the CoD (by training with the full data set and computing the error on the same training). This would have two effects: it would decrease computation time and would yield an optimistic (high-biased) estimate of the CoD [13, 14].

There are $_mC_k = m!/k!(m-k)!$ predictor combinations for k predictor genes out of m total genes. For t target genes where $1 \le t \le m$, there are $tm!/k!(m-k)!$ coefficients to be calculated. Since incremental relations between smaller and larger predictor sets are important, it is necessary to calculate the CoD for k predictor gene combinations, for each k of 1, 2, 3, ..., to some stopping point. A large storage space is required for all or part of the CoD results.

3. PREDICTION SYSTEM DESIGN

The next two sections discuss design issues of the codetermination algorithm. Although a perceptron-based prediction system [6-8] was also implemented, the design of a logic-filter-based prediction system will be the focus of this paper.

3.1. Ternary Logic Filter

A ternary logic filter is a non-linear predictor that has an k-gene input and an output. The input is ternary data for the k genes, and the output is the predicted value of the quantized target expression. Rather than using the ternary-quantized conditional expectation for the predictor, the conditional mode is used. This requires less computation, the differences between it and the conditional expectation are small, and it avoids predicting values that have not occurred in the samples, a desirable property for coarse quantization. The ternary logic filter is defined by a logic table constructed via the conditional probability of the output Y given input data X as follows:

$$Y = \Psi(\mathbf{X}) = \begin{cases} -1 & \text{if } P(Y=-1|\mathbf{X}) \text{ is highest} \\ 0 & \text{if } P(Y=0|\mathbf{X}) \text{ is highest} \\ 1 & \text{if } P(Y=1|\mathbf{X}) \text{ is highest} \end{cases} \qquad (3)$$

Filter design falls to computing the conditional probability for each input-output pair, ($\mathbf{x}$, y). For any observation vector $\mathbf{x}$, $\Psi(\mathbf{x})$ is the value of Y seen most often with $\mathbf{x}$ in the sample data. The size of table defining the predictor grows exponentially with the number of predictor variables, and the number of conditional probabilities to estimate increases accordingly. For two input variables and ternary data, there are $3^2 = 9$ conditional probabilities to estimate; for three variables, there are $3^3 = 27$.

For gene expression ratio data, the number of input vectors available for filter design is very limited; in fact, we often do not observe all vectors. When applying the filter to test data, there may be inputs not observed during design. The quantized expected value of Y, $T(E[Y])$, is used as the output from the filter for all input vectors that are not observed in the training.

3.2. Adjusting CoD Values

As additional inputs are added to the filter, the information content of the input increases. The ability to predict the target gene expression value increases, and ε_n decreases. In the worst case, when the additional gene carries no information, ε_n remains constant. This is not observed in the designed filters. Statistical fluctuations caused by the estimation error resulting from a small number of samples allow logic filters designed with more inputs to have a larger ε_n and smaller CoD value than the same filter without one of the inputs. After CoD values are calculated for gene combinations with more than one input value, the results are compared with the CoD values from all gene combination with one less input. If the new CoD value is less than the CoD value calculated with the smaller number of inputs, than the CoD value is assigned to the CoD value from the smaller input set. This assures that the estimated CoD increases as information content increases.

4. PARALLEL ANALYSIS OF GENE EXPRESSION (PAGE)

PAGE is a parallel implementation of the codetermination algorithm discussed above. A system to analyze microarray data is depicted in Fig. 1. The 3 main modules of PAGE generate all k-predictor gene combinations for a target T, calculate the CoD, and adjust the CoD. For any target T, all one-gene predictor combinations are determined. The predictor combinations are the passed into the CoD calculator which builds the logic filters and calculates a CoD value for each combination. The values are sorted and CoD histograms are created. For the one-gene predictor, there is no CoD adjustment. The one-gene inputs are now finished and the 2-gene inputs begin. All 2-gene predictor combinations are determined, and CoD values for each predictor combination is calculated. Results are sorted and histograms are created. At this point, CoD values calculated from 2-gene combinations and from 1-gene combinations exist. The CoD values from the 2-gene combination are adjusted using the CoD values from the 1-gene combinations. The adjusted results are sorted and adjusted histograms are created. The CoD values from the 2-gene combination are again compared to the CoD values from the 1-gene combinations. This time, instead of adjusting the CoD values, instances where the CoD values from 2-gene combinations is less than the CoD values from the 1-gene combination are deleted from the results. These results are considered redundant because all the information is carried by the 1-gene predictor, and the addition of the second gene is unnecessary. The result that are not redundant are sorted and histograms are created. The 3-gene combinations follow the same procedure as the 2-gene combination: all combinations are determined, CoD values are calculated, and the results are adjusted, and a redundancy check is performed. The results from all stages are sorted and histograms are created. The procedure could continue for larger combinations, but in practice the number of combinations make the calculations impractical.

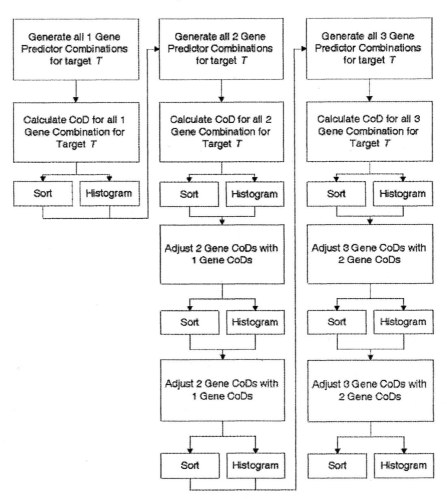

Fig. 1 A schematic of the PAGE system.

4.1. Generating Predictor Combinations

For any target T, all one-gene predictor combinations are determined. The number of combinations is $_{m-1}C_k = (m-1)!/(m-k-1)!k!$ where m is the number of genes in the microarray data, and k is the number of genes in the predictor. The value of $m-1$ arises from the fact that the target is not allowed to be a predictor. Since the predictor combinations exclude the target gene, the list of the predictors needs to be recreated for every target.

4.2. Calculating CoD Values

For each predictor gene combination a CoD value is calculated. The number of gene combinations becomes very large quickly, so the CoD calculation is performed in parallel on a Beowulf-type cluster of workstations using MPI [15] and is shown schematically in Fig. 2. The combinations are partitioned evenly across p processors. Each processor then randomly partitions the sample from the microarray data into a training and test set. The training set develops the logic filter, and the test set is used to get a CoD value. The training and testing is repeated many times (128 times for our studies) estimate the CoD. The average of the CoD values is reported. After each processor has completed its share of the work, the CoD values are written to a file on a shared disk.

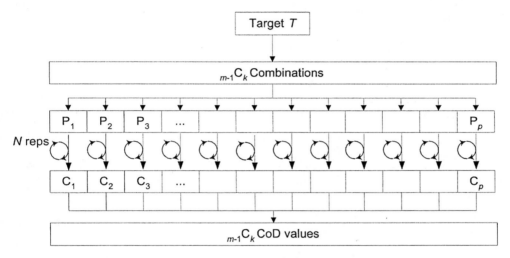

Fig 2. A schematic of the CoD calculation algorithm. For each target, the combinations are partitioned across the processors. The CoD is calculated N times for each combination.

The parallel performance of the code as measured by the parallel efficiency is shown in Table 1. The code was run on the Biowulf cluster at the Center for Information Technology of the National Institutes of Health using AMD 1.4 GHz XP/Athlon processors with 1 Gbyte of memory. CoD values for all 3-gene predictor combinations from one target of a data set consisting of 31 samples of 587 genes were calculated on 1, 2, 4, 8, 16 and 32 processors (the data set will be described in the Applications of PAGE Results section). Most of the inefficiencies in the code can be related to the amount of I/O being performed. In the serial case, the results are written directly to the global disk, in the parallel case, writing to the global disk causes disk contention problems that increase with the number of processors. This issue was partially alleviated by writing to a local disk, and then copying from local to the global disk. Some of the overhead involved in the recopying data is hidden by small load imbalances.

**Table 1. Parallel performance of the CoD calculation for a single gene target
with 3-input genes and 33,366,840 predictor gene combinations**

Number of Processors	Runtime (hours)	Efficiency
1	55.94	100%
2	33.90	83%
4	17.00	82%
8	8.55	82%
16	4.30	81%
32	2.29	76%

4.3. Adjusting CoD Values

As mentioned earlier the CoD values are compared against CoD values from predictor combinations with less inputs. As an example, assuming that the calculated CoD value using gene G_1, G_2, and G_3 is smaller than the CoD value calculated using G_1 and G_3, then the adjusted CoD value for combination (G_1, G_2, G_3) would be set to the CoD value of combination (G_1, G_3). The same algorithm is used when looking for redundant data. The only difference is whether or not the results are kept. The 2-gene adjusted results are kept to simplify adjusting the 3-gene results.

The algorithm used to adjust the data is also performed in parallel and is shown schematically in Fig. 3. The CoD values are partitioned equal across p processors. All CoD values from the smaller set S are read into an array in memory. A lookup table is created for the CoD values in the array with the first combination that starts with 0, 1, 2, ..., m-k. Given any combination of predictors, the CoD value is compared to the CoD values for all combination subsets with 1 less gene. The lookup table facilitate moving around the array in memory. The array indices for the subset are not known and the array must be searched for subsets. There are bounds that can be set to limit the search space, but no closed expression for the indices has been found.

The parallel performance of the adjustment code is shown in Table 2. The code was run on the same AMD 1.4 GHz XP/Athlon processors as the CoD calculations. The output of the one of the 3-gene predictor CoD benchmark was adjusted using 2-gene CoD results for the same target. The code was run on 1, 2, 4, 8, 16, and 32 nodes. Even though it is parallel performance is low, on the Biowulf cluster nodes are already allocated by the PAGE system, so instead of leaving them idle we run with all the nodes allocated. Of course when running with more than 32 processors, care should be taken to make sure that the adjust time does not increase with additional processors. This case was not seen with 2-gene predictors, which has less work/processor and is it is unlikely the addition for more processors would show this behavior. Nevertheless, results may vary when extrapolating far beyond the results in Table 2.

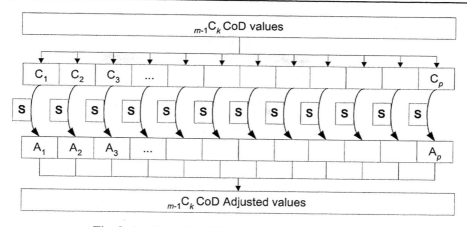

**Fig. 3. A schematic of the adjusting algorithm. CoD
values are partitioned across N processors, and compared
to the CoD values from the predictors with one less input gene.**

**Table 2. Parallel performance of the CoD adjustment for a 3-input genes, 33,366,840
predictor gene combinations results with a 2-input gene, 171,405 combination result.**

Number of Processors	Runtime (hours)	Efficiency
1	0.51	100%
2	0.36	70%
4	0.20	65%
8	0.12	50%
16	0.08	39%
32	0.06	25%

5. APPLICATIONS OF PAGE RESULTS

While the CoD analysis mainly focuses on finding individual connections among genes, using this information in the context of a biological problem remains a huge challenge and an open problem. A recent case study, currently being published [16], is a good example of how CoD results can be used.

The study investigates whether the network of interactions that regulate gene expression can be modeled by existing mathematical techniques. Studies of the ability to predict a gene's state based on the states of other genes, such as CoD analysis, suggest that it may be possible to abstract sufficient information to build models of the system that retain some characteristics of the real system.

As an attempt to determine whether certain biological behavior could be captured in a Markov chain model, a small network based on microarray data observations of a human cancer, *melanoma*, was built and simulated by a Markov chain. A Markov chain is a widespread statistical methodology to enable estimation of complex models via simulation, in particular, in the context of biological system. This requires developing criteria to select a small set of genes from which to build a Markov chain and developing a method to construct transition rules from microarray data. Since it would be unrealistic to study all genes in one regulatory network because of the limit on computational resource and the perception that biological system seems to be composed of small, function-centered regulatory sub-networks, a small set of genes was chosen based on the following criteria: (1) predictive relationships based on CoD analysis using parallel system described above, (2) available biological prior-knowledge.

The set of data used in the study are the results from a study of 31 melanoma samples [17]. For that study, total messenger RNA was isolated directly from melanoma biopsies, and fluorescent cDNA from the message was prepared and hybridized to a microarray containing probes for 8,150 cDNA (representing 6,971 unique genes). Preprocessing the microarray data produced a data set of 587 genes from the melanoma samples. The data set was then fed into PAGE for CoD analysis, as described previously. PAGE performed the codetermination analysis of melanoma expression data in two months of computer time using 4 sets of 32 processors in a Beowulf type cluster of workstations [18] at the Center for Information Technology, National Institutes of Health (NIH). This analysis would have required about four years on 4 sequential computing machines with similar processor speeds. The processors actually used in the analysis were significantly slower than the processors used for the benchmarking (Intel Pentium III 450/550 MHz). On average, it took about 6.5 hours to perform all the steps in Fig 1. Using the fastest processors cuts this time to about 3 hours. The parallel system generated hundreds of millions of CoD values for this study to provide fundamental bottom-line information. Using the CoD information and the criteria above, we identified 10 genes as our targets for the network study.

The focus of the study was the steady-state behavior of the Markov chain constructed from the multivariate relationships and the transition rules, both estimated from the data. Ones interested in detail results should refer to the original paper [16]. We will only briefly present a result (not published in the original paper). One of the most pleasing outcomes from the analysis is that there exists only very small numbers of states with very high probability of being in the steady state. We can eventually identify attractors of the system and their basins. Fig. 4 shows one of the attractors and its basins found in the study. In the figures, we show eight different states within the attractor. Only 5 of 10 genes in the system change their states in the transitions within the attractor. Also, those attracting states are very close to the observation made through microarray experiments, which is consistent with biological reasoning.

6. Conclusions

A system to calculate estimates of CoD values for large-scale cDNA microarray experiments has been developed. PAGE uses parallel processing to perform the calculation on a reasonable time scale. The availability of CoD results for all predictor combinations allows for a high level study of information flow driving gene regulation. As an example, a Markov chain study used CoD results to produce a network that has properties similar to those found in biological systems.

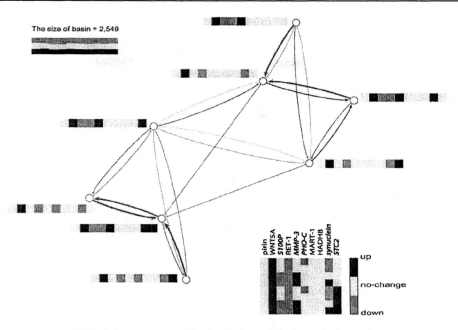

Fig. 4 Attractor and its basin from Markov chain study

ACKNOWLEDGEMENT

All biological data used in our experiments came from the Cancer Genetics Branch at the National Human Genome Research Institute. This study utilized the high-performance computational capabilities of the Biowulf/LoBoS3 cluster at the National Institutes of Health, Bethesda, Md.

REFERENCES

[1]   DeRisi, J. L., Iyer, V. R., and Brown, P. O. (1997). Exploring the metabolic and genetic control of gene expression on a genomic scale. *Science* **278**, 680-6.

[2]   Schena, M., Shalon, D., Heller, R., Chai, A., Brown, P. O., and Davis, R. W. (1996). Parallel human genome analysis: microarray-based expression monitoring of 1000 genes. *Proc. National Academy Science U S A* **93**, 10614-9.

[3]   Evan, G., and Littlewood, T. (1998). A matter of life and cell death. *Science* **281**, 1317-22.

[4]   McAdams, H. H., and Shapiro, L. (1995). Circuit simulation of genetic networks. *Science* **269**, 650-6.

[5]   Yuh, C. H., Bolouri, H., and Davidson, E. H. (1998). Genomic cis-regulatory logic: experimental and computational analysis of a sea urchin gene. *Science* **279**, 1896-902.

[6]   Kim, S., Dougherty, E. R., Bittner, M. L., Chen, Y., Sivakumar, K., Meltzer, P., and Trent, J. M., 2000. A general nonlinear framework for the analysis of gene interaction via multivariate expression arrays. *J. Biomed. Optics 5*, 411-424.

[7]   Kim, S., Dougherty, E. R., Chen, Y., Sivakumar, K., Meltzer, P., Trent, J. M., and Bittner, M. 2000. Multivariate measurement of gene expression relationships. *Genomics 67*, 201-209.

[8]   Dougherty, E. R., Kim, S., and Chen, Y., 2000. Coefficient of determination in nonlinear signal processing. *Signal Processing 80*, 2219-2235.

[9]   Shmulevich, I., Dougherty, E. R., Kim, S., and W. Zhang, "Probabilistic Boolean Networks: A Rule-based Uncertainty Model for Gene Regulatory Networks," *Bioinformatics*, 18, 261-274, 2002.

[10] Kauffman, S. A., "Metabolic Stability and Epigenesis in Randomly Constructed Genetic Networks," *Theoretical Biology*, 22, 437-467, 1969.

[11] Kauffman S. A., *The Origins of Order: Self-organization and Selection in Evolution* (Oxford Univ. Press, New York) 1993.

[12] Chen, Y., Dougherty, E. R., and Bittner, M. L. 1997. Ratio-based decisions and the quantitative analysis of cDNA microarray images. *J. Biomed. Optics 2*, 364-374.

[13] Dougherty, E. R., "Small Sample Issues for Microarray-Based Classification," *Comparative and Functional Genomics*, 2, 28-34, 2001.

[14] Devroye, L., Gyorfi, L., Lugosi, G., *A Probabilistic Theory of Pattern Recognition*, Springer-Verlag, New York, 1996.

[15] Gropp, W., Lusk, E., and Skjellum, A. 1999. *Using MPI: Portable Parallel Programming with the Message Passing Interface*, The MIT Press, Cambridge, MA.

[16] Kim, S., Li H., Dougherty E.R., Cao N., Chen Y., Bittner M. L., and Suh, E. B., Can Markov Chain Models Mimic Biological Regulation? *Journal of Biological Systems*, 2002 (accepted)

[17] Bittner M. L., Meltzer P., Chen Y., Jiang Y., Seftor E., Hendrix M., Radmacher M., Simon R., Yakhini Z., Ben-Dor A., Sampas N., Dougherty E. R., Wang E., Marincola F., Gooden C., Lueders J., Glatfelter A., Pollock P., Carpten J., Gillanders E., Leja D., Dietrich K., Beaudry C., Berens M., Alberts D., Sondak V., Hayward N. and Trent J., Molecular classification of cutaneous malignant melanoma by gene expression profiling. *Nature* 406 (2000) pp. 536-540.

[18] Sterling, T. L., Salmon, J., Becker, D. J., and Savarese, D. F. 1999. *How to Build a Beowulf: A guide to the Implementation and Application of PC Clusters,* The MIT Press, Cambridge, MA.

In: Biocomputing
Editor: Phillip A. Laplante, pp. 29-41

ISBN 1-59033-889-8
2003 © Nova Science Publishers, Inc.

Chapter 3

VIRTUAL TISSUE ENGINEERING OF CARDIAC MUSCLE: COMPUTATIONAL ASPECTS

Richard H. Clayton and Arun V. Holden

Computational Biology Laboratory, School of Biomedical Sciences
University of Leeds
Leeds, LS2 9JT
United Kingdom
arun@cbiol.leeds.ac.uk http://www.cbiol.leeds.ac.uk

ABSTRACT

Cardiac arrhythmias remain an important cause of morbidity and mortality in the Western world. Although the underlying mechanisms can be studied experimentally, investigations are in general limited to mapping electrical activity on the heart surface. Computational models of the heart provide a way to study the mechanisms, onset and control of arrhythmias. We illustrate a framework for cardiac virtual tissue engineering in which the geometry, electrophysiology and regional properties of heart tissue can be specified. Application of cardiac virtual tissues in arrhythmia research, the design and pre-screening of anti-arrhythmics, and defibrillation technologies is already raising high performance computing and visualization demands that require reformulation of current planned Grid technologies.

Key-Words: - ion channel, action potential, virtual tissue, excitable cell, ordinary differential equations, partial differential equations

1. INTRODUCTION

Functional genomics - the prediction of the effects of an identified gene mutation on the behavior of cells, tissues, organs and systems, and even on the survival of the organism, raises major conceptual and practical problems. However, it is not just an academic study - not only are the genetic bases for some pathologies being identified, but they are also providing insights into physiological mechanisms that allow novel interventions. The heart provides a testbed for developing computational functional genomics, as its function as a pump is well understood, and the heart is composed of well characterised cells that interact locally. Cardiac virtual tissue engineering is also an area of considerable practical importance, as sudden cardiac death (occurring within one hour of a cardiac event) accounts for about 50% of the 300,000 *per annum* deaths from cardiovascular disease in the U.S. [1]. Computational reconstruction of arrhythmias, from clinical data streams, anatomy, and the results from animal experiments, provides a means for evaluating mechanisms [2], developing new intervention strategies [3], and pre-screening antiarrhythmics drugs [4]. They provide one component in an integrated strategy that aims to reduce the incidence of sudden cardiac death by an order of magnitude in little over a decade.

The rhythm of the heart is primarily determined by the dynamics of a few dozen proteins involved in ion transfer, sequestration and binding in cardiac cells. Cardiac cell electrical activity can be modelled by a stiff, high order, system of ordinary differential equations, where the components are empirically obtained from experiments on expressed channels, membrane patches, single cells, and isolated tissue, and the overall model is validated against cell electrophysiology. Cell models, combined with histology (the distribution of cell type, and their orientation) and anatomy form virtual cardiac tissues. The effects of changed protein kinetics (produced by mutations, pharmacological agents or pathological processes) on whole heart function can be computed. The mechanical beating of the heart is triggered by propagating waves of excitation, the action potentials. The ionic mechanisms that generate the action potential of different cardiac muscle cell types have been described quantitatively in terms of the kinetics of transmembrane proteins - the voltage-dependent channels, ion exchangers and pumps. Changes in the kinetics of ion transport and sequestration mechanisms may be produced by the actions of drugs, or by mutations; and their effects on the electrical activity of the cell computed. Cardiac cells are electrically coupled to their neighbours by gap junctions, and so propagation in cardiac tissue can be modelled either by a lattice of linearly coupled cells, or as a continuum, by a reaction diffusion equation in which the diffusion coefficient codes the cell-to-cell coupling. Cardiac cells are cylindrical, and organised in sheets, with adjacent fibres having a similar orientation. Propagation is faster in the direction of the fiber axis, as the longitudinal cell-to-cell coupling is larger than the transverse coupling: this anisotropy can be represented by a diffusion tensor. Thus local histology - fibre orientation, and the sheet structure of cardiac muscle - can be coded into the anatomical geometry of the atrial and ventricular cardiac chambers [5].

Ventricular fibrillation (VF) is an often lethal cardiac arrhythmia that is initiated and sustained by mechanisms that remain poorly understood. VF is difficult to study in patients because it is characterised by a sudden onset and it is life threatening unless terminated quickly. In experimental studies the spread of electrical activity on the surface of the heart during VF can be mapped either using an array of electrodes or by perfusing the tissue with a dye that has voltage dependent fluorescent properties. Using these techniques the complex spatiotemporal patterns of activation during VF have been observed and quantified. Most evidence supports the idea that electrical activation during VF is sustained by re-entry, during which a wave of electrical activity continuously circulates to activate tissue that has recovered from the previous activation [6].

The properties of re-entry have been studied using theoretical (from the physics of nonlinear wave processes in excitable media) and computational approaches, as well as by experiment. In the last decade, these insights have begun to make an impact on the study of cardiac arrhythmias. By comparing surface activation patterns produced by virtual tissues with those seen experimentally, candidate mechanisms can be quantitatively evaluated. Cardiac virtual tissues provide a tool for unifying clinical and experimental data on the molecular, cellular and tissue behaviors of the heart in health and disease, that enables the development of new understanding and new therapeutic approaches to arrhythmias.

2. PROBLEM FORMULATION

Virtual tissue engineering can be considered in terms of how experimental data is quantitatively incorporated to construct virtual tissues; how the virtual tissues are validated; and how their applications in basic and clinical research raises computational problems.

2.1. Model Construction

Action potentials V(t) from different parts of the heart - the sinoatrial node, the atria, the atrio-ventricular node, the Purkinje fibers, and the ventricles - have different characteristics, in terms of their maximum rates of rise and fall, duration and shape, that change as the interval between action potentials alters. However, they are all generated by similar processes. The lipid membrane separates solutions of different ionic composition, and contains ion-selective channels, exchangers and pumps. Ionic current flow through these proteins charges or discharges the membrane capacitance, and for a unit area of cell membrane

$$C_m dV/dt = -I_{ionic}, \tag{1}$$

where C_m is the specific membrane capacitance (1 $\mu F.cm^{-2}$) and I_{ionic} the membrane current density, which is the sum of the current densities for the different, ion-selective channels, pumps and exchangers.

Differences in action potentials from different parts of the heart result from quantitative differences in the expression of the different ion transport proteins. Different cardiac cells have different densities of different channels - a preliminary catalogue of cardiac channel types lists channels in terms of their properties and expressed equivalents [9]. Figure 1 A illustrates the component ionic currents for a recent model of a sinoatrial node cell; the sum of all the currents would form the right hand side of equation 1, and each component current could be described by differential equations with a few dynamic variables.

Ions flowing into and out of the cell can produce concentration changes in restricted spaces, and Ca^{++} inside the cell may be bound to proteins (such as calmodulin), trigger Ca^{++} induced Ca^{++} release or be sequestered in the sarcoplasmic reticulum. Figure 1(B) illustrates a simplified schematic for intracellular Ca dynamics in a cardiac cell.

Combining the equations for transmembrane ion flows and intracellular sequestration and binding processes produces a virtual cell as a nonlinear system of differential equations.

These equations are typically high order, and stiff (time scales vary from fractions of a ms to a few s) and may be numerically solved on a simple single workstation. They may have equilibrium, multiple equilibrium, periodic, quasi-periodic and even chaotic solutions, and some care may be needed in numerical integration or path following algorithms, for obtaining solutions and bifurcation curves [10].

In a virtual cardiac tissue the changes in cell properties produced by differential channel expression can be represented as a gradient in parameter values for the cell excitation models, and adjacent cells can be resistively coupled in a coupled ordinary differential equation (CODE) lattice model, or as a spatially heterogeneous partial differential equation model.

A simple one dimensional model, produced by coupling cells together into a strand can be used to predict the effects on some tissue behaviours of changed membrane protein kinetics produced by drugs or mutations. The resistive coupling between cells can be directly input from experimentally obtained cell-to-cell coupling conductance, or selected to give an appropriate conduction velocity (given the cell size), or spatially modulated by an immunochemical visualisation of connexin density.

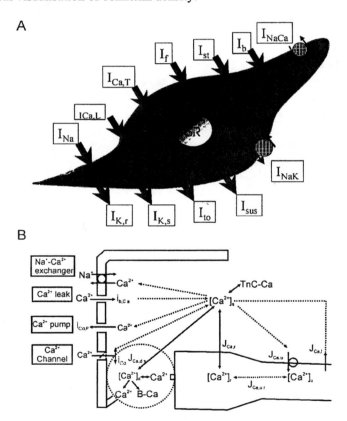

Figure 1. (A) Diagram of ionic currents in a model of a single sinoatrial cell [7]. For clarity, the depolarising currents are on the upper surface, and the repolarsing currents on the lower surface. Each current represents the dynamics of a density ty of specific membrane proteins. distrubuted over the membrane surface. (B) Schematic diagram of spatially distributed intracellular Ca^{++} dynamics, of entry, CICR, binding and sequestration [8]

For idealised one dimensional models it is reasonable to model each cell separately, as a one dimensional CODE lattice, as to span the ventricular wall one would only need a few hundred cells. However, it is computationally more efficient to represent the string of resistively coupled cells by a continuous, one dimensional excitable medium represented by a partial differential equation

$$\partial V/\partial t = D\ \partial^2 V/\ \partial x^2 - I_{ionic}\ /C_m, \tag{2}$$

as the space step necessary for integration is larger than the size of a cardiac cell. D is a diffusion coefficient, with the dimensions of cm^2/s, and may vary with space. Such one dimensional virtual tissues have be used to compute

- the threshold for the initiation and propagation velocity of solitary action potential,
- the mutual annihilation of action potentials by collision,
- rate dependent changes in action potential duration and conduction velocity, and
- the vulnerability of the tissue to re-entrant excitation [5].

Although early computations with biophysically detailed equations of propagation in two dimensions used a CODE lattice approach they used supercomputer resources such as the Connection Machine [11]. The continuum approach, using numerical solution of the partial differential equation, is far more practical, even though the short time steps (needed for the rapid upswing of the atrial and ventricular action potentials) and the associated small space step needed for stable computations do require a fast processor.

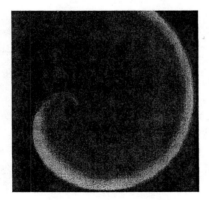

Figure 2. Spiral wave solution of a 6 cm square of ventricular virtual tissue: the position of the spiral can be defined by the position of its tip, and its orientation by a phase. The voltage variable is coded by the grey scale (white depolarised).

The generalisation of a solitary wave and a wave train in a 2-dimensional medium is a plane wave and a plane wave train. Since cardiac cells are cylindrical, with a higher density of gap junctions (and stronger electrical and mechanical coupling) at their ends, and organised in sheets, propagation in cardiac tissue is anisotropic, being faster in directions parallel to the fibre axis. Thus for 2-dimensional virtual tissues the diffusion coefficient of (2) is replaced by a vector, determined by the fibre orientation.

The normal wavefront velocity v of a wavefront depends on its curvature k

$$v = v_0 - Dk \tag{3}$$

where D is the effective diffusion coefficient. The rate and velocity dependence of propagation velocity allows spiral wave solutions that provide a model for re-entry in a homogenous, isotropic medium. Spiral waves rotate around central area of conduction block, and can be characterised by their period of rotation, size of central core, and movement of the tip. A spiral wave rotates as fast as it can, providing a tachyarrhythmic source about 10 times faster than the normal heart rate. At a specific instant in time the spiral wave has a location (given by the position of its tip) and a spatial orientation or phase. The tip can totate rigidly around a circular core, or meander biperiodically [5].

In 3-dimensional media re-entrant waves rotate around a thin, rod-like filament and the motion of filaments in three-dimensional media is more complex than simple meander in two-dimensional media. Even in a homogeneous, isotropic medium it is possible for a simple untwisted scroll to be unstable, even though the two-dimensional spiral wave in a medium with the same parameters is stable. Thus, although some questions may be answered using 1- and 2-dimensional media some problems need to be explored using 3-dimensional tissues or even in anatomically accurate whole heart geometries. In the heart wall, transmural rotational anisotropy can lead to the breakup of a transmural scroll wave. Numerical experiments have shown that transmural rotational anisotropy, differences in action potential duration, and the maximum steepness of the action potential duration dependence on rate all contribute to destabilisation of transmural waves.

The heart has a moving three-dimensional geometry and a complicated architecture, with the ventricular wall made up of layers of muscle fibres that are locally parallel, but whose orientation changes smoothly with position over the heart surface. The fibre orientation changes through 120 degrees on going through the ventricular wall. In ventricular tissue, cells are aligned parallel to each other in fibres which are wrapped around the ventricles in a spiral pattern that accounts for the twist that is observed during contraction. Action potentials are conducted more quickly along cardiac fibres than across them, so fibre orientation is important for modelling anisotropic action potential propagation. Thus the diffusion coefficient of (1) is replaced by a tensor, that can be obtained from histological measurements of fibre orientation [12], or diffusion tensor magnetic resonance imaging of the heart [13].

Detailed human ventricular anatomy is not currently available in the public domain. The canine ventricular anatomy and fiber orientation data set from the University of Auckland [12], is available for academic use (http://www.bioeng.auckland.ac.nz), as is a data set for the rabbit ventricles [14] from UCSD (http://cmrg.ucsd.edu). There is a clear need for data sets human ventricular anatomy that include fiber orientation, and a number of groups are working to develop these. Human hearts vary in size and shape however, and a library of normal and pathological human ventricular geometry would be a tremendously valuable resource for modeling studies.

2.2. Model Validation

Cell models can be validated by comparison of their solutions - action potential properties, intracellular Ca^{++} transients - against intracellular recordings from isolated cells, *in vitro* tissue or even monophasic action potentials recorded from the surface of the (human) heart by suction electrodes. In practice, models for cells from laboratory animals are better validated than human cell models. Figure 3 illustrates the activity of virtual cells - action potentials in a standard (wild type) guinea-pig, epicardial ventricular cell model, and for a model in which the kinetics for the mutant I_{Kr} channels of human Long QT syndrome (LQT2), and Na^+ channel of human Long QT syndrome (LQT3) has been inserted. The channel kinetics were obtained by single channel recordings from frog oocytes expressing wild-type or mutant-channels.

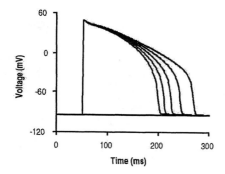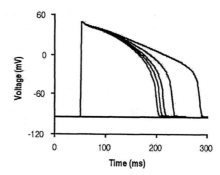

Figure 3. Action potentials, triggered from the resting state, for wild -type (shortest duration action potential of (A) and (B), and (A) increasing LQT2 and (B) increasing LQT3 expression [15]

Although virtual cells can be validated by direct comparison with intracellular recordings, all real *in vivo* or *in vitro* cardiac tissue is heterogenous, anisotropic, and 3-dimensional. 1- and 2-dimensional virtual tissues are idealisations, that allow the quantitative importance of different phenomena (channel kinetics, action potential restitution, heterogeneities, anisotropies) to be evaluated. The rotation frequency of spiral waves, and how it "ages"with a time scales of seconds can be compared to the frequencies of tachycardias [15]. Quasi- 1- or 2- dimnsional preparations can be dissected from cardiac tissue, for example, as thin epicardial slices approximating a rectangular 2-dimensional preparation, or frezing the inside of the ventricles to leave a thin prolate hemispherical surface preparation. However, the main validation of the 1- and 2- dimensional virtual tissues follows from the interpretation of their results. As an example, for a two dimensional model of guinea-pig epicardial tissue, incorporation of the changed channel kinetics for human LQT1, LQT2 and LQT3 allows the computation of re-entry characteristics for these arrhythmias in a chimaeric virtual tissue. Clinically, episodes of ventricular tachycardia in LQT3 syndrome are four times more likely to be lethal than similar episodes in LQT1 syndrome; this correlates with the interpretation of the computed behavior of spiral wave solutions, where the LQT1 spirals meander over twice as large an area as the LQT3 spirals. The greater the meander of the spiral, the more likely is for it to reach an inexcitable boundary of the heart, where the re-entry maybe extinguised and the arrhythmia self-terminate [4,16].

3-dimensional virtual tissue slabs, which can be homogeneous or heterogeneous; isotropic, uniformly anisotropic or have rotational anisotropy, are also idealizations that allow mechanisms to be explored. Comparison with experimental results is in terms of qualitative phenomenology, where 3-dimensional behavior in the virtual tissue slabs is used to explain 2-dimensional movies of electrical activity seen on the cardiac surfaces. As an example, the observation of different patterns of spatio-temporal activity on the endocardial and epicardial surfaces, and the organization of this activity in domains that have dominant frequencies [17] has been modeled by simple virtual tissue slabs [18,19] and the same quantitative measures of complexity (Karhunen-Loève decomposition) applied to real and virtual spatio-temporal data streams [20].

Whole ventricle models can be validated by experiments on isolated, perfused hearts where the spatio-temporal pattern of activity on the surface can be visualized by voltage sensitive dyes [6]; experiments on *in situ* hearts, where activity recorded by an array of electrodes on the heart surface can be related to body surface recordings [21]. Clinical recording of activity at several points on the endocardial surface can be used to construct an endocardial activation map of surface activity, but most clinical recordings are from the body surface as multi-channel electrocardiograms, or body surface maps. Embedding a virtual heart within a torso allows the reconstruction of body surface maps, and provides a link between virtual tissue engineering and noninvasive clinical recordings [22]. All the components exist for this link to made into a closed loop, where virtual tissue engineering is used to interpret clinical recordings.

2.3. Applications

Virtual cardiac tissues were originally constructed as a research project that aimed to provide a constructive framework for the integrative physiology of the heart, linking molecular, intracellular, membrane, cellular and tissue behaviours in a quantitative, predictive hierarchy of computational models [23]. Virtual cardiac cells have been used since the mid 1960s; the availability of poweful laboratory workstations, and access to supercomputing resources in the 1990s has allowed the development of virtual tissues, as specific case studies.

Now virtual tissues for various parts of the normal heart of different species have been developed and validated, they are being applied as routine laboratory tools. Such applications produces a quantitative shift, from customised to mass throughput, that is being accelerated by the introduction of high throughput, quantitative techniques into experimental biology. This is producing an increasing demand for high performance computational resources. A simple but approximate unit for quantifying the computational load is the Euler heartbeat; the number of floating point operations that would be necessary to compute one heartbeat. The resting human heart rate is about 70/min., so a heartbeat lasts about 1s. Computing 1s of electrical activity, with a time step of 10s and a space step of 0.1mm would require some 10^{14} floating point operations using fixed time and space steps. In practice, for a normal heartbeat this could be reduced by a few orders of magnitude by using variable time and space steps. However, re-entrant arryhthmias are as fast as they can be, and efficient adaptive grid methods offer no more than an order of magnitude reduction in CPU time. A further order of magnitude reduction in CPU time can be achieved by using look-up tables rather than evaluating the voltage dependent kinetics. The addition of contraction [24] and even intracardiac fluid dynamics [25] increases the demand so a complete computation of excitation, contraction and blood ejection during a single heart beat remains an unachieved grand challenge project.

Further additions to cardiac excitation models - intracellular Ca^{++} dynamics as a spatio-temporal (sparks, oscillations, and waves) rather than a purely temporal process; coupling ionic fluxes to metabolism and second messenger systems; and functional remodeling due to activity produced up and down regulation of membrane proteins - all produce a continuing inflation in the computational exchange rate for an Euler heart beat. As a rough unit, illustration of an arrhythmia may take a few such Euler heart beats; systematic investigation of the effects of one drug on such an example may take 10^3 Euler heart beats. There is an ever increasing demand for compute performance, as virtual tissues are applied to pharmacological prescreening [4] and defibrillation methods [2,3,26]. As virtual tissues are applied to real clinical problems there will be occasional needs for real time solution.

3. PROBLEM SOLUTION

We developed our virtual cardiac tissues on single processor laboratory workstations, as running a particular illustration for several CPU weeks was an acceptable solution. For many applications, virtual cell or 1-dimensional virtual tissues are adequate, and some problems (like how the vulnerability to re-entry is altered by changes in channel properties) can be reduced to 1-dimensional tissues. For essentially 3-dimensional problems where the anatomy rather than details of excitation are of prime interest, simplified models (coupled map lattices, rather than partial differential equations) can be adequate [27]. Coupling such simple models for the spread of excitation with detailed models of local excitation provides a computational efficient approach that allows virtual tissue computations on a workstation [28].

Since the summer of 2001 our laboratory has had a 0.15 Teraflop cluster grid that is 24/7 devoted to virtual tissue engineering, of one 24-processor SMP machine and a 52-processor grid, all running under Sun GridEngine, and 3-dimensional computations are parallelized. Multiple runs of 2- and 3-dimensional slab computations, as illustrated in Fig. 4, are run on the grid, while whole ventricle computations, as illustrated in Fig 5, are run on the SMP machine. Cardiac virtual tissues are not suitable for parallelization over a large number of thin nodes, but many problems may be run in batch mode over an assembly of separate processors.This provides a massive increase in throughput, as it is production line rather than developmental scientific computation, and has produced a visulisation bottleneck. This is being solved by grid- enablement of visualisation tools, to allow computational guidance of multiple runs and computational steering of whole ventricle computations.

Even for small slabs of virtual tissue, computational guidance requires extracting of aspects of the solution from the full 3-dimensions+time computational output, that can be fully explored in virtual reality through a vrml browser. This is illustrated in Fig 4, that displays surface views and filaments from frames from a movie of activity in a slab.

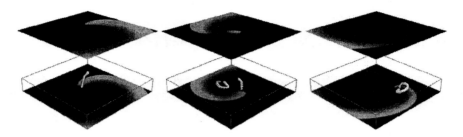

Figure 4 Evolution of a transmural scroll wave: surface view and intramural filament. Successive frames from movie illustrating breakdown due to transmural differences in action potential duration

For the canine ventricular geometry illustrated in Fig. 5, 1 s of activity could be simulated in about 3.5 hours with OpenMP parallel computation on 8 750 MHz Sun Ultrasparc III processors. For currently funded research projects in our laboratory - virtual prescreening, mapping the pacemaker of the heart andon the mechanisms of ventricular fibrillation - a sustained 24/7 compute demand of 1-2 teraflop is anticipated within 3 years. In the UK alone there are several other centres with similar research requirements.

4. CONCLUSION

A local high performance cluster grid can provide the sustained compute resource for the experimental applications of cardiac virtual tissues, and GridEngine has provided an adequate management tool. A cluster grid also provides the medico-legal security necessary for dealing with clinical data, and the commercial security for dealing with virtual prescreening of pharmaceutics. Anticipated demand from virtual tissue engineering is too large to be met by any planned global grid structures; what we envisage is a small number of high performance cluster grids in different laboratories, linked into a secure network for data and virtual tissue exchange *i.e.* a knowledge network of cluster grids.

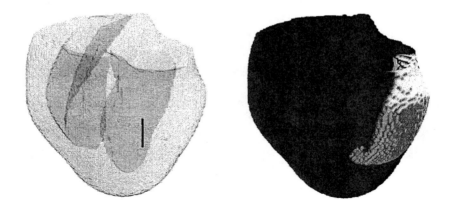

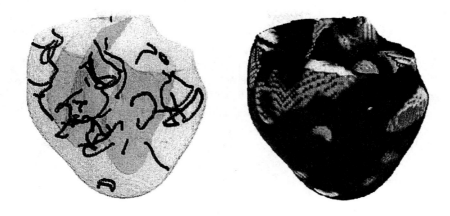

Figure 5. Filaments (left) and surface activity during initial scroll wave (top) and
fully developed fibrillation in virtual canine ventricle.

ACKNOWLEDGEMENTS

Richard Clayton is funded by British Heart Foundation Lectureship BS 98/001. We are also grateful to the MRC and EPSRC for additional funding

REFERENCES

[1]   R.J.Myerburg, Scientific gaps in the prediction and prevention of sudden cardiac death, *Journal of Cardiovascular Electrophysiology* Vol 13 No. 7, 2002, pp. 709- 723.

[2]   V.N.Biktashev, & A.V.Holden, Reentrant arrhythmias and their control in models of mammalian cardiac tissue. *Journal of Electrocardiology* Vol. 32 Supplement, 1999, pp.76-83.

[3]   V.N.Biktashev, A.V.Holden & H.Zhang, A model for the action of external current onto excitable tissue. *International Journal of Bifurcation and Chaos* Vol. 7, 1997, pp. 477-485

[4]   V.N.Biktashev & A.V.Holden, Enhanced self-termination of re-entrant arrhythmias as a pharmacological strategy for antiarrhythmic action, *Chaos,* Vol. 12, 2002, pp. 843-851.

[5]   A.V.Panfilov & A.V.Holden, (eds.) *The Computational Biology of the Heart*, John Wiley & Sons, Chichester. 1997

[6]   R.A.Gray, A.M.Pertsov & J.Jalife, Spatial and temporal organization during cardiac fibrillation *Nature* Vol.**392** 1988 pp. 75-8.

[7]   H.Zhang, A.V.Holden, I. Kodama, H.Honjo, M.Lei, T.Varghese, M.R.Boyett, Mathematical models of action potentials in the periphery and center of the rabbit sinoatrial node. *American Journal of Physiology (Heart and Circulatory Physiology)* Vol. 279, 2000, pp. H397-H421.

[8]   M.R.Boyett, H.Zhang, A.Garny, A.V.Holden, Control of pacemaker of the heart, by intracellular Ca^{++}. *Philosophical Transactions of the Royal Society of London.* B Vol. 359, 2001, pp. 1091-1110.

[9]   M.R.Boyett, S.M.Harrison, N.C.Janvier, S.O.McMorn, J.M.Owen & Z.Shui, A list of vertebrate cardiac ionic currents. Nomenclature, properties, function and cloned equivalents. *Cardiovascular Research* Vol.32, 1996, pp. 455-481.

[10] A.Varghese & R.L.Winslow, Dynamics of the calcium subsystem in cardiac Purkinje fibres. *Physica-D,* Vol.68, pp.354-386.

[11] R.L.Winslow, A.Kimball, A.Varghese, D.Noble, Simulating cardiac sinus and atrial network dynamics on the Connection Machine. *Physica D* Vol. 64, 1993, pp. 281-298

[12] P.M.F.Nielsen, I.J.E.LeGrice, B.H.Smaill & P.J.Hunter, Mathematical model of geometry and fibrous structure of the heart *American Journal of Physiology (Heart and Circulatory Physiology)* Vol. 29, 1991, pp. H1365-H78.

[13] D.F.Scollan, A.Holmes, J.Zhang & R.L.Winslow, Reconstruction of cardiac ventricular geometry and fibre orientation using magnetic resonance imaging *Annals of Biomedical Engineering* Vol. 28, 2000, pp. 934-44.

[14] F.J.Vetter & A.D.McCulloch, Three-dimensional analysis of regional cardiac function: a model of rabbit ventricular anatomy. *Progress in Biophysics and Molecular Biology,* Vol. 69, 1998, pp.157-183.

[15] A.V.Holden & H.Zhang, Characteristics of atrial re-entry and meander computed from a model of a single atrial cell. *Journal of Theoretical Biology Vol.* 175, 1995, pp.545-551.

[16] R.H.Clayton, A.Bailey, V.N.Biktashev & A.V.Holden, Re-entry in computational models of LQT1, LQT2 and LQT3 myocardium. *Journal of theoretical Biology* 208, 2001, pp. 215-225.

[17] A.V.Zaitsev, O.Berenfeld, J.Jalife, S.F.Mironov & A.M.Pertsov, Distribution of excitation frequencies on the epicardial and endocardial surfaces of fibrillating ventricular walls of the sheep heart. *Circulation Research Vol.* 86, 2000, pp.408-417.

[18] V.N.Biktashev, A.V.Holden, S.F.Mironov, A.M.Pertsov & A.V.Zaitsev, Three-dimensional aspects of re-entry in experimental and numerical models of ventricular fibrillation *International Journal of Bifurcation and Chaos* Vol. 9, 1999, 695-704

[19] V.N.Biktashev, A.V.Holden, S.F.Mironov, A.M.Pertsov & A.V.Zaitsev,On two mechanism for the domain structure of ventricular fibrillation. *International Journal of Bifurcation and Chaos* Vol. 11, 2001, pp. 1035-1064.

[20] V.N.Biktashev & A.V.Holden, Characterization of patterned irregularity in locally interacting, spatially extended systems. *Chaos Vol.* 11 2001, pp. 653-664.

[21] M.P.Nash, C.P.Bradley, A.Kardos, A.J.Pullan & D.J.Paterson, An experimental model to correlate simultaneous body surface and epicardial electropotential recordings. *Chaos, Solitons and Fractals* Vol. 13, 2002, pp. 1735-1742.

[22] R.H.Clayton & A.V.Holden, Computational framework for simulating the mechanisms and ECG of re-entrant ventricular fibrillation. *Physiological Measurement* Vol. 23, 2002, pp.707-726.

[23] D.Noble, Modeling the heart - from genes to cells to whole organ. *Science* Vol. 295 2002, pp1678-1682.

[24] M.P.Nash, P.J.Hunter, Computational mechanics of the heart - from tissue structure to ventricular function. *Journal of Elasticity* Vol. 61, 2000, pp.113-141.

[25] S.J.Kovacs, D.M.McQueen & C.S.Peskin, Modeling cardiac fluid dynamics and diastolic function. *Philosophical transactions of the Royal Society of London A* Vol. 359, 2001, 1299-1314.

[26] I.R.Efimov, Virtual electrodes in virtual reality of defibrillation, *Journal of Cardiovascular Electrophysiology Vol.* 13 No. 7, 2002, pp.680-1.

[27] A.V.Holden, M.J.Poole, J.V.Tucker, An algorithmic model of the mammalian heart: propagation, vulnerability, re-entry and fibrillation, *International Journal of Bifurcation and Chaos* Vol. 6, 1996, pp. 1623-1635

[28] M.J.Poole, A.V.Holden, & J.V.Tucker, Hierarchical reconstructions of cardiac tissue, *Chaos, Solitons and Fractals Vol.* 13 No. 8, 2002, pp. 1581-1612.

In: Biocomputing
Editor: Phillip A. Laplante, pp. 43-53

ISBN 1-59033-889-8
2003 © Nova Science Publishers, Inc.

Chapter 4

COMBINING PARTICLE SWARMS AND *K*-NEAREST NEIGHBORS FOR THE DEVELOPMENT OF QUANTITATIVE STRUCTURE-ACTIVITY RELATIONSHIPS

Walter Cedeño and Dimitris K. Agrafiotis
3-Dimensional Pharmaceuticals, Inc.
665 Stockton Drive, Exton, Pennsylvania 19341
USA
(Walter.Cedeno, Dimitris.Agrafiotis)@3dp.com

ABSTRACT

The development of quantitative structure-activity relationship (QSAR) models for computer-assisted drug design is an established technique in the pharmaceutical industry. QSAR models provide a framework for predicting a compound's biological activity based on its chemical structure or properties, and can significantly reduce the time to discover a new drug. In this paper we describe the application of a new optimization technique, particle swarms, to develop QSAR models based on *k*-nearest neighbor and kernel regression. Particle swarms is a population-based stochastic method that has been used successfully for feature selection. Each individual explores the feature space guided by its previous success and that of its neighbors. Success is measured by the predictivity of the resulting model as determined by *k*-nearest neighbor and kernel regression. The swarm flies through the feature space in search of the global minimum, guided by the regression error. The technique is evaluated using well-known QSAR data sets and compared to other machine learning techniques.

Key-Words: Computer-assisted drug design, QSAR, feature selection, particle swarm, case-based reasoning, *k*-nearest neighbors, kernel regression, optimization, stochastic methods.

1. INTRODUCTION

The increasing amount of information available in digital form has prompted the development of novel data mining methodologies that can help scientists process and interpret large volumes of data faster and with greater reliability. Artificial intelligence methods, such as artificial neural networks (ANN) [1], classification and regression trees (CART) [2], and k-nearest neighbor classifiers (KNN) [3], have been used extensively for this purpose [4, 5]. These methods are used in drug design to correlate some measure of biological activity with a set of physicochemical, structural and/or electronic parameters, known as descriptors, of the compounds under investigation. It is assumed that the biological activity of a compound is related to its chemical structure, and can therefore be inferred from a carefully chosen set of molecular descriptors. The key challenge is to determine which set of descriptors correlates best with biological activity. Since it is not possible to know in advance which molecular features are most relevant to the problem at hand, a comprehensive set of descriptors is usually employed, chosen based on experience, software availability, and computational cost.

However, is well known, both in the chemical and statistical fields, that the number of features used in a QSAR model can greatly affect its accuracy. The presence of noise and irrelevant or redundant features can cause the method to learn the idiosyncrasies of the individual samples and lose sight of the broad picture that is essential for generalization beyond the training set [6]. This problem is compounded when the number of observations is also relatively small, as is often the case in molecular design. If the number of variables is comparable to the number of training patterns, the parameters of the model may become unstable and unlikely to replicate if the study were to be repeated. The large number of descriptors available also increases the risk of chance correlations [29]. Comparing the results against randomly generated results is commonly used to verify that the risk of chance correlations is low or does not exist.

Feature selection is often used to remedy this situation and improve the accuracy of a classification or regression technique. Feature selection works by identifying a small subset of necessary and sufficient features that can be used as input to the underlying predictor. Feature selection algorithms can be divided into three main categories [7]: 1) those where the selection is embedded within the basic regression algorithm, 2) those that use feature selection as a filter prior to regression, and 3) those that use feature selection as a wrapper around the regression. The latter has a long history in statistics and pattern recognition, and is the method of choice for QSAR.

Feature selection can be viewed as a heuristic search, where each state in the search space represents a particular subset of the available features. In all but the simplest cases, an exhaustive search of the state space is impractical, since it involves n!/(n-r)!r! possible combinations, where n is the total number of available features and r is the number of features selected. Several search algorithms have been applied to this problem, ranging from simple greedy approaches such as forward selection or backward elimination [8], to more elaborate methodologies such as simulated annealing [9], evolutionary programming [10], genetic algorithms [11, 12, 13, 14], artificial ants [15], and more recently binary particle swarms [16].

In this paper, we present a new algorithm that combines particle swarms [17], an approach inspired from the study of human sociality, with k-nearest neighbor and kernel regression [18, 19]. KNN is a simple regression technique based on the principle that similar structures tend to exhibit similar biological properties. Particle swarms explore the feature space through a population of individuals, which adapt by returning towards previously successful regions. Their movement is stochastic, and is influenced by the individuals' own memories as well as that of their peers. This method, which was originally intended for searching multidimensional continuous spaces, is adapted to the problem of feature selection by viewing the location vectors of the particles as feature weights. The fitness, a measure of success, for each individual is based on the regression error calculated using k-nearest neighbor and kernel regression and the observed values. In the remaining sections, we provide the key algorithmic details of particle swarm optimization and kernel regression, and discuss its advantages compared to alternative methods based on simulated annealing and artificial neural networks using three classical data sets from the QSAR literature.

2. METHODS

k-Nearest Neighbor and Kernel Regression

The k nearest neighbor method (KNN) is an intuitive method used extensively for classification. Given a pattern to classify, KNN works by selecting the k most similar patterns from a set of well-known classified data (training data) and choosing the class with the most representatives in the set. Similarity between elements is typically measured using the Euclidean distance in some appropriate feature space or some other suitable metric. KNN is a lazy algorithm, i.e., it defers data processing until needed. The algorithm uses local information and adapts well to changes in the training data. Two main drawbacks of KNN are its susceptibility to noise and the curse of dimensionality. These can be alleviated using normalization and feature weighting to calculate the distance between patterns according to the equation

$$d(p,q) = \sqrt{\sum_{i=1}^{n} w_i (p_i - q_i)^2} \qquad (1)$$

where n is the number of features, p_i and q_i are the i-th feature values for patterns p and q respectively, and w_i the weight for the i-th feature. The implementation of KNN used in this work is based on k-d trees in order to reduce the computational cost associated with calculating distances [27].

Kernel regression is a closely related non-parametric methodology that uses local information to obtain a prediction. The main difference from KNN is that the prediction for q is based on the weighted average of the response values of all the patterns p in the neighborhood of q. Usually, the kernel function gives more weight to points that are closer to q.

In this work, the patterns are chemical structures and the input features are a suitably chosen set of molecular descriptors. We use KNN to select the k most similar structures to a structure q and apply kernel regression using the kernel function

$$f(p,q) = \frac{1}{1 + d(p,q)} \qquad (2)$$

where $d(p,q)$ is the KNN distance as given in equation 1. The prediction for q is obtained by

$$y'(q) = \frac{\sum\limits_{i=i}^{k} y_i f(p_i, q)}{\sum\limits_{i=i}^{k} f(p_i, q)} \tag{3}$$

where k is the number of nearest neighbors selected for q, p_i is the i-th nearest neighbor of q, y_i is the known response value of p_i, and $f(p, q)$ the kernel function given in equation 2.

Particle Swarms

Particle swarms (PS) is a relatively new optimization paradigm introduced by Kennedy and Eberhart [17]. The method is based on the observation that social interaction, which is believed to play a crucial role in human cognition, can serve as a valuable heuristic in identifying optimal solutions to difficult optimization problems. Particle swarms explore the search space using a population of individuals, each with an individual, initially random, location and velocity vector. The particles then "fly" over the state space, remembering the best solution encountered. Fitness is determined by an application-specific objective function $f(x)$. During each iteration, the velocity of each particle is adjusted based on its momentum and the influence of the best solutions encountered by itself and its neighbors. The particle then moves to a new position, and the process is repeated for a prescribed number of iterations. In the original PS implementation [20], the trajectory of each particle is governed by the equations:

$$v_i(t+1) = v_i(t) + \eta_1 \cdot r \cdot (p_i - x_i(t)) + \eta_2 \cdot r \cdot (p_{b(i)} - x_i(t)) \tag{4}$$

and

$$x_i(t+1) = x_i(t) + v_i(t+1) \tag{5}$$

where x_i and v_i are the current position and velocity of the i-th particle, p_i is the position of the best state visited by the i-th particle, $b(i)$ is the particle with the best fitness in the neighborhood of i, and t is the iteration number. The parameters η_1 and η_2 are called the cognitive and social learning rates, and determine the relative influence of the memory of the individual versus that of its neighborhood. In the psychological metaphor, the cognitive term represents the tendency of organisms to repeat past behaviors that have proven successful or have been reinforced by their environment, whereas the social term represents the tendency to emulate the successes of others, which is fundamental to human sociality. In effect, these terms introduce a tendency to sample regions of space that have demonstrated promise. r is a random number whose upper limit is a constant parameter of the system, and is used to introduce a stochastic element in the search process.

Kennedy defined four models of PS [20]. The full model, which places equal influence to the cognitive and social influence, the social-only model, which involves no cognitive learning, the cognitive-only model, which has no social component, and the selfless model, which is a social-only model in which the individual is excluded from consideration in determining its neighborhood's best. The neighborhood represents a subset of the population surrounding a particular particle. The neighborhood size defines the extent of social interaction, and can range from the entire population to a small number of neighbors on either side of the particle (i.e. for the i-th particle, a neighborhood size of 3 would represent particles i-1, i, and i+1).

The present work employs Shi and Eberhart's [21, 22] variant of the PS algorithm, which makes use of an inertia weight, ω, to dampen the velocities during the course of the simulation, and allow the swarm to converge with greater precision:

$$v_i(t+1) = \omega \cdot v_i(t) + \eta_1 \cdot r \cdot (p_i - x_i(t)) + \eta_2 \cdot r \cdot (p_{b(i)} - x_i(t)) \tag{6}$$

Larger values of ω induce larger transitions and thus enable global exploration, whereas lower values facilitate local exploration and fine-tuning of the current search area.

The location vector for each particle corresponds to the feature weights used in KNN. The swarm searches for the best subset of features and corresponding weights that minimize the regression error in the training data. All weights are limited to values in the range [0, 1]. Location vectors that move outside this range are normalized at run time. In the present study, only a fixed number of features with the highest weight values were used to construct the actual models.

Data Sets

The methods were tested on three well-known data sets: antifilarial activity of antimycin analogues (AMA) [8, 23, 10, 11, 12], binding affinities of ligands to benzodiazepine/GABAA receptors (BZ) [24, 25], and inhibition of dihydrofolate reductase by pyrimidines (PYR [26]). These data sets have been the subject of extensive QSAR studies, and have served as a test bed for many feature selection algorithms. To allow comparison with previous neural network-based approaches, the number of input features were taken from the literature. These details are summarized in Table 1. In all three cases, the descriptor data were normalized to [0, 1] prior to modeling.

Table 1. Data Set Size and Model Parameters Used

Data Set	Number of Samples	Number of Features	Features Selected	Hidden Neurons
AMA	31	53	3	3
BZ	57	42	6	2
PYR	74	27	6	2

Implementation

All programs were implemented in the C++ programming language and are part of the DirectedDiversity® software suite [28]. All calculations were carried out on a Dell Inspiron 8000 laptop computer equipped with a 1 GHz Pentium IV Intel processor running Windows 2000 Professional.

3. RESULTS AND DISCUSSION

Our study was designed to accomplish two main goals. The first was to establish a reasonable set of parameters for KNN and PS, and the second was to determine whether the method offers any advantages over other commonly used learners and search algorithms. In particular, we compare the results with those obtained from a recent study of the relative performance of simulated annealing (SA), particle swarms (PS), and artificial ant colonies (AA) using artificial neural networks (ANN) [16].

Following common practice, two measures were used to define the quality of the resulting models. The first is the training correlation coefficient R

$$R = \frac{N \sum_{i=1}^{N} y_i \tilde{y}_i - \sum_{i=1}^{N} y_i \sum_{i=1}^{N} \tilde{y}_i}{\sqrt{\left[N \sum_{i=1}^{N} y_i^2 - \left(\sum_{i=1}^{N} y_i \right)^2 \right]\left[N \sum_{i=1}^{N} \tilde{y}_i^2 - \left(\sum_{i=1}^{N} \tilde{y}_i \right)^2 \right]}}, \tag{8}$$

and the second is the cross-validated correlation coefficient, R_{CV}, resulting from leave-one-out (LOO) cross-validation. The latter is obtained by systematically removing one of the patterns from the training set, building a model with the remaining cases, and predicting the activity of the removed case using the optimized weights. This is done for each pattern in the training set, and the resulting predictions are compared to the measured activities to determine their degree of correlation. Since KNN does not involve any training and the patterns in the test data do not participate in the prediction of their own response values, the training R corresponds to the LOO cross-validated R_{CV}.

Since ANNs are unstable predictors, the learning (R) and generalization (R_{CV}) performance of the best ANN models discovered by each optimization technique were established by running several training and cross-validation experiments, each starting with a different random seed, and averaging the respective statistics (50 runs were used for computing R, and 10 for computing R_{CV}). The same procedure was followed for all three data sets under investigation. A summary of the results for the ANN work can be found in Table 2.

**Table 2. Top Models Selected by the Binary Particle Swarms (PS),
Simulated Annealing (SA), and Artificial Ant Algorithm (AA)**

Data Set	ANN-PS Selected Variables	$\mu(R_{CV})$	$\sigma(R_{CV})$	ANN-SA Selected Variables	$\mu(R_{CV})$	$\sigma(R_{CV})$	ANN-AA Selected Variables	$\mu(R_{CV})$	$\sigma(R_{CV})$
AMA	31,35,49	0.831	0.009	31,37,49	0.837	0.008	31,37,49	0.838	0.010
BZ	1,4,6,9,20,21	0.900	0.019	1,4,5,9,20,23	0.901	0.005	0,1,5,9,11,14	0.890	0.009
PYR	0,5,10,16,19,22	0.808	0.014	1,2,3,5,19,22	0.795	0.015	1,2,3,5,19,22	0.796	0.005

Comparing the results obtained with ANN models (Table 2) to our approach using KNN with PS optimization is not straightforward. The method described here not only selects the best features for the kernel regression model, but also identifies their relative contribution. The search space associated with this problem is more complex than the binary problem of classical feature selection. To ensure a fair comparison, each optimization run was constrained to 1000 fitness function evaluations. The best models obtained using KNN with PS are shown in Table 3.

**Table 3. Top Models Selected by *k*-Nearest Neighbor
Kernel Regression and Particle Swarms (KNN-PS)**

Data Set	KNN-PS Selected Variables	Variable Weights	K	$\mu(R_{CV})$
AMA	18,19,49	0.834, 0.981, 0.975	1	0.861
BZ	1,5,9,14,20,22	0.689, 0.818, 0.812, 0.727, 0.651, 1.000	4	0.858
PYR	0,4,10,11,17,19	0.769, 0.869, 0.853, 0.625, 0.997, 0.811	3	0.829

Comparing the results in Tables 2 and 3 (summarized in Figure 1) shows that KNN-PS outperformed the other methods for the AMA and PYR data sets. KNN-PS did not do too well for the BZ data set. This might be due, in part, to our choice of kernel function. A different kernel function might do a better job with the BZ data set. Although these results do not allow any definitive conclusions to be drawn, they demonstrate the potential of KNN-PS for building QSAR models with good generalization power. Moreover, the method provides the researcher with a weight vector that can help in determining the relative importance of each feature. One can argue that the weight of the selected features leads to an easier interpretation of the model, which, in some cases, may be a significant advantage over ANN. On the other hand, KNN-based methods are computationally more intensive, since they require a search for the k nearest neighbors before calculating the predicted value. On the other hand, since there is no training involved, KNN regression provides an easier way to include new data in the model.

The implementation of KNN-PS that we used allowed the selection of models with a predefined minimum and maximum numbers of features. Only features with weights above a given threshold (up to the predefined maximum) were chosen to be part of the model. Given this flexibility, we performed some experiments where we allowed solutions containing between 1 and 10 features with a threshold of 0.1. Although the results are not shown, models with up to 10 features produced slightly better LOO R_{CV} values. More experiments must be conducted in order to determine conclusively if the results are due to the ability of KNN-PS to do feature weighting.

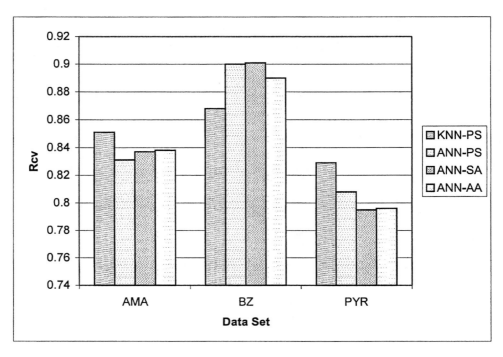

Figure 1: LOO cross-validation values for the best models discovered by KNN-PS, ANN-PS, ANN-SA, and ANN-AA for the AMA, BZ, and PYR data sets. For ANN-PS, ANN-SA, and ANN-AA, the value represents the average over 10 independent cross-validation runs.

We also investigated the effect of k, the neighborhood size in KNN, on the performance of KNN-PS. For these, we varied k between 1 and 10 and kept all other parameters fixed for each data set. The results are illustrated in Figure 2. In general, a neighborhood size between 1 and 4 produced better models than larger values, which is consistent with prior experience. Small values for k are usually better, but the exact value depends on the data set.

The results presented here provide a glimpse of the potential of using KNN-PS for QSAR modeling. The success demonstrated in the present study provides motivation for the exploration of further improvements of the technique. One obvious variation is to replace our simplistic kernel with a more sophisticated function. Another possibility is to improve the ability of particle swarms to create niches and converge to multiple solutions during the same run. This will improve the ability of the technique to detect and avoid local minima and hopefully lead to better QSAR models.

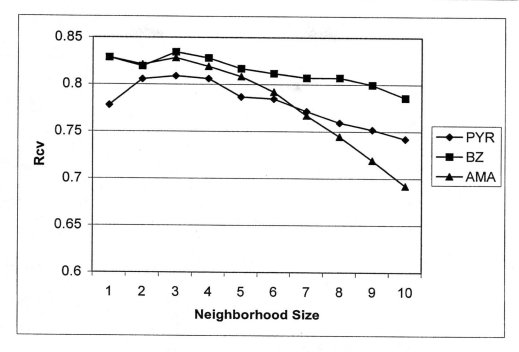

**Figure 2: Effect of neighborhood size (*k*) on the
generalization performance of KNN-PS.**

4. Conclusion

A promising new algorithm for feature selection for QSAR has been described. The
method is based on a combination of *k*-nearest neighbors and kernel regression with particle
swarms, an intrinsically parallel optimization paradigm inspired from the study of human
sociality. In two out of three data sets that were tested, the method was able to identify QSAR
models with better generalization power as measured by LOO cross-validation. The two main
advantages of KNN-PS over ANN are the ability to add more data to the model without
retraining, and the ability to infer the relative importance of the selected features based on
their respective weights which leads to greater interpretability. Further empirical work is
necessary in order to fully assess the effects of the various parameters on the algorithm.
Finally, as in most published QSAR studies, the number and variety of data sets tested are
small, and only through extensive investigation with additional real-world applications can
we measure the robustness of the algorithm.

Acknowledgments

The author would like to thank Dr. Victor S. Lobanov, Michael Farnum, Dmitrii N.
Rassokhin, Huafeng Xu, and Sergei Izrailev of the Research informatics group at 3-

Dimensionsl Pharmaceuticals, Inc., for many useful discussions, and Dr. Raymond F. Salemme for his insightful comments and support of this work.

REFERENCES

[1]   Devillers, J., Ed., *Neural networks in QSAR and drug design*, Academic Press, 1996.

[2]   Breiman, L., Friedman, J. H., Olshen, R. A. and Stone, C. J., *Classification and Regression Trees*, Wadsworth Int. Group, Belmont, California, USA, 1984.

[3]   Parzen, E, On the Estimation of a Probability Density Function and Mode, *Ann. Math. Stat.*, Vol. 33, 1962, pp. 1065-1076.

[4]   van de Waterbeemd, H., Ed., Chemometric methods in molecular design, in *Methods and Principles in Medicinal Chemistry*, Vol. 2, VCH, Weinheim, 1995.

[5]   Hansch, L., and Leo, C., Exploring QSAR. *Fundamentals and applications in chemistry and biology*, American Chemical Society, Washington, DC, 1996.

[6]   Manallack, D. T., Ellis, D. D., and Livingston, D. J., Analysis of linear and nonlinear QSAR data using neural networks, *J. Med. Chem.*, Vol. 37, 1994, pp. 3758-3767.

[7]   John, G., Kohavi, R., and Pfleger, J., Irrelevant features and the subset selection problem, in Machine learning: *proceedings of the 11-th international conference*, Morgan-Kaufmann, 1994, pp. 121-129.

[8]   Selwood, D. L., Livingstone, D. J., Comley, J. C. W., O'Dowd, A. B., Hudson, A. T., Jackson, P., Jandu, K. S., Rose, V. S. and Stables, J. N., Structure-activity relationships of antifilarial antimycin analogues, a multivariate pattern recognition Study, *J. Med. Chem.*, Vol. 33, 1990, pp. 136-142.

[9]   Sutter, J. M., Dixon, S. L., and Jurs, P. C., Automated descriptor selection for quantitative structure-activity relationships using generalized simulated annealing, *J. Chem. Info. Comput. Sci.*, Vol. 35, 1995, pp. 77-84.

[10] Luke, B. T., Evolutionary programming applied to the development of quantitative structure-activity relationships and quantitative structure-property relationships, *J. Chem. Info. Comput. Sci.*, Vol. 34, 1994, pp. 1279-1287.

[11] Rogers, D. R., and Hopfinger, A. J., Application of genetic function approximation to quantitative structure-activity relationships and quantitative structure-property relationships, *J. Chem. Info. Comput. Sci.*, Vol. 34, 1994, pp. 854-866.

[12] So, S., and Karplus, M., Evolutionary optimization in quantitative structure-activity relationship: an application of genetic neural networks, *J. Med. Chem.*, Vol. 39, 1996, pp. 1521-1530.

[13] Yasri, A., and Hartsough, D., Toward an optimal procedure for variable selection and QSAR model building, *J. Chem. Info. Comput. Sci.*, Vol. 41, 2001, pp. 1218-1227.

[14] Hasegawa, K., Miyashita, Y., and Funatsu, K., GA strategy for variable selection in QSAR studies: GA-based PLS analysis of calcium channel antagonists, *J. Chem. Info. Comput. Sci.*, Vol. 37, 1997, pp. 306-310.

[15] Izrailev, S., and Agrafiotis, D. K., Variable selection for QSAR by artificial ant colony systems, *SAR and QSAR in Environ. Res.*, Vol. 13, No.3-4, 2002, pp. 417-423.

[16] D. K. Agrafiotis, and W. Cedeño, Feature selection for structure-activity correlation using binary particle swarms, *J. Med. Chem.*, Vol. 45, 2002, pp. 1098-1107

[17] Kennedy, J., and Eberhart, R. C., Particle swarm optimization, *Proc. IEEE International Conference on Neural Networks*, Perth, Australia, IEEE Service Center, Piscataway, NJ, IV, 1995, pp. 1942-1948.

[18] Nadaraya, E.A., On estimating regression. *Theory of Probability and its Applications*, Vol. 9, 1964, pp. 141-142.

[19] Watson, G.S., Smooth regression analysis, *Sankhya - The Indian Journal of Statistics*, Series A, Vol. 26, pp. 359-372.

[20] Kennedy, J., The particle swarm: social adaptation of knowledge, *IEEE International Conference on Evolutionary Computation*, Indianapolis, IN, IEEE Service Center, Piscataway, NJ, 1997, pp. 303-308.

[21] Shi, Y. H., and Eberhart, R. C., A modified particle swarm optimizer, *IEEE International Conference on Evolutionary Computation*, Anchorage, AL, 1998.

[22] Shi, Y. H., and Eberhart, R. C., Parameter selection in particle swarm optimization, *7-th Annual Conference on Evolutionary Programming*, San Diego, CA, 1998.

[23] Wikel, J. H., and Dow, E. R., The use of neural networks for variable selection in QSAR, *Bioorg. Med. Chem. Lett.*, Vol. 3, 1993, pp. 645-651.

[24] D. J. Maddalena, G. A. R. Johnson, Prediction of receptor properties and binding affinity of ligands to benzodiazepine/GABAA receptors using artificial neural networks, *J. Med. Chem.*, Vol. 38, 1995, pp. 715-724.

[25] So, S. S., and Karplus, M., Genetic neural networks for quantitative structure-activity relationships: improvements and application to benzodiazepine affinity for benzodiazepine/GABAA receptors, *J. Med. Chem.*, Vol. 39, 1996, pp. 5246-5256.

[26] Hirst, J. D., King, R. D., and Sternberg, M. J. E., *J. Comp.-Aided Mol. Design*, Vol. 8, 1994, pp. 405-420.

[27] D. K. Agrafiotis, and V. S. Lobanov, "An efficient implementation of distance-based diversity metrics based on k-d trees", *J. Chem. Info. Comp. Sci.*, Vol. 39, 1999, pp. 51-58.

[28] D. K. Agrafiotis, R. Bone, R. F. Salemme and R. Soll. "System and method for automatically generating chemical compounds with desired properties". *US Patent 5,463,564*, October 31, 1995.

[29] J. G. Topliss and R. P. Edwards, "Chance factors in studies of quantitative structure-activity relationships", *J. Med. Chem.*, Vol. 22, 1979, pp. 1238-1244.

In: Biocomputing
Editor: Phillip A. Laplante, pp. 55-71

ISBN 1-59033-889-8
2003 © Nova Science Publishers, Inc.

Chapter 5

TOWARDS MINIMAL ADDITION-SUBTRACTION CHAINS USING GENETIC ALGORITHMS

Nadia Nedjah and Luiza de Macedo Mourelle
Department of Systems Engineering and Computation,
Faculty of Engineering,
State University of Rio de Janeiro, Rio de Janeiro,
Brazil
(nadia | ldmm)@eng.uerj.br   http://www.eng.uerj.br/~ldmm

ABSTRACT

Addition and addition-subtraction chains consist of a sequence of integers that allow one to efficiently compute power T^E, where T varies but E is constant. The shorter the addition (addition-subtraction) chain is, the more efficient the computation. Solving the optimisation problem that yields the shortest addition (addition-subtraction) is *NP*-hard. There exists some heuristics that attempt to obtain reduced addition (addition-subtraction) chain. We obtain better addition (addition-subtraction) chains than those yield by so far known heuristics using genetic algorithms. The addition (addition-subtraction) chains are nearly minimal.

Key-Words:- Addition-subtraction chains, Exponentiation, Cryptography.

1. INTRODUCTION

The modular exponentiation is a common operation for scrambling and is used by several public-key cryptosystems, such as the RSA encryption scheme [20] and DSS for digital signature [17]. It consists of a repetition of modular multiplications: $C = T^E \bmod M$, where T is the plain text such that $0 \leq T < M$ and C is the cipher text or vice-versa, e is either the public or the private key depending on whether T is the plain or the cipher text, and M is called the modulus. The decryption/encryption operations are performed using the same procedure, i.e. using the modular exponentiation, which itself is a sequence of modular multiplications.

Recently, several cryptosystems based on the Abelian group defined over elliptic curves were proposed. In these cryptosystems, the inverse of an element is easily obtained. Hence for such groups one can compute exponentiations by an interleaved sequence of multiplications and divisions.

The performance of existing cryptosystems is primarily determined by the implementation efficiency of the modular multiplication/division. As the operands are usually large (i.e. 1024 bits or more), and in order to improve time requirements of the encryption/decryption operations, it is essential to attempt to minimise the number of modular multiplications/divisions performed.

A simple procedure to compute $C = T^E \bmod M$ is based on the paper-and-pencil method. This method requires $E-1$ modular multiplications. It computes all powers of T: $T \rightarrow T^2 \rightarrow T^3$ $\rightarrow ... \rightarrow T^{E-1} \rightarrow T^E$, where $\rightarrow$ denotes a multiplication. The paper-and-pencil method computes more multiplications than necessary. For instance, to compute T^{31}, it needs 30 multiplications. However, T^{31} can be computed using only 7 multiplications: $T \rightarrow T^2 \rightarrow T^3 \rightarrow$ $T^5 \rightarrow T^{10} \rightarrow T^{11} \rightarrow T^{21} \rightarrow T^{31}$. But if division is allowed, T^{31} can be computed using only 5 multiplications and 1 division: $T \rightarrow T^2 \rightarrow T^4 \rightarrow T^8 \rightarrow T^{16} \rightarrow T^{32} \rightarrow^- T^{31}$, with $\rightarrow^-$ denotes a division.

The basic question is: what is the fewest number of multiplications and/or divisions to compute T^E, given that the only operation allowed is multiplying or dividing two already computed powers of T? Answering the above question is NP-hard, but there are several efficient algorithms that can find a near optimal ones. In the rest of the paper, addition chains will be considered as a particular case of an addition subtraction chain.

Evolutionary algorithms are computer-based solving systems, which use evolutionary computational models as key element in their design and implementation. A variety of evolutionary algorithms have been proposed. The major ones are *genetic algorithms* [9]. They have a conceptual base of simulating the evolution of individual structures via the Darwinian natural selection process. The process depends on the performance of the individual structures as defined by its environment. Genetic algorithms are well suited to provide an efficient solution of NP-hard problems [3].

In the rest of this paper, we will present a novel method based on the addition-subtraction chain method that attempts to minimise the number of modular multiplications and divisions, necessary to compute exponentiations. It does so using genetic algorithms.

This paper will be structured as follows: Section 2 presents the most cited addition-subtraction chain based methods; Section 3 gives an overview on genetic algorithms concepts; Section 4 explains how these concepts can be used to compute a minimal addition-subtraction chain. Section 5 presents some useful results.

2. THE ADDITION-SUBTRACTION CHAIN BASED METHODS

The addition-subtraction chain based methods attempt to find a chain of numbers such that the first number of the chain is 1 and the last is the exponent E, and in which each member of the chain is the sum/difference of two previous members. A formal definition of an addition chain is as in Definition 1:

Definition 1: An *addition-subtraction chain* of length l for an integer n is a sequence of integers $(a_0, a_1, a_2, ..., a_l)$ such that $a_0 = 1$, $a_l = n$ and $a_k = a_i \pm a_j$, $0 \le i \le j < k \le l$.

For the addition-only chain based methods the exponent is represented by the binary representation while for the addition-subtraction based methods, the exponent is recoded, generally using the Booth algorithm. There are several recoding rules and some of them are exposed and formally analysed in [19]. Recoding of the exponents should normally reduce the number of multiplications and divisions that are necessary to compute exponentiation. Generally speaking, the recoding techniques use the fact that:

$$2^{i+j-1} + 2^{i+j-2} + 2^{i+j-3} + \ldots + 2^i = 2^{i+j} - 2^I$$

to collapse a block of ones in order to obtain a minimal sparse representation. Using the notation $\bar{1} = -1$ to represent signed binary numbers, the signed canonical binary representation of $23_{10} = 10111 = 10\bar{1}00\bar{1}$. Note that the binary representation contains four digits different from 0 while the recoded representation contains only three such digits. However, recoding methods need an efficient way to compute the inverse of elements, as it is the case with cryptosystems that are based on elliptic curves where the inverse is easily obtained.

Finding a minimal addition-subtraction chain for a given number is *NP*-hard [4]. Therefore, heuristics are used to attempt to approach such a chain. The most used heuristic consists of scanning the digits of E from the less significant to the most significant digit and grouping them in partitions P_i. The size of the partitions can be constant or variable [8], [10], [11]. Modular exponentiation methods based on constant-size partitioning of the exponent are usually called (*recoding*) *m-ary methods*, where m is a power of two and $\log_2 m$ is the size if a partition and modular exponentiation methods based on variable-size are usually called (*recoding*) *window methods*. There exist several strategies to partition the exponent. The generic computation of such methods is given in Fig. 1, where E^* is the recoded exponent, V_i and L_i denote the value and length of partition P_i respectively, $\wp(E)$ denotes the partition set of exponent E and b the number of such partitions.

Algorithm (Recoding) ModularExpo of Fig. 1 is inspired from the following observation:

$$T^E = T^{V_{b-1} \times 2^{(b-1)L_{b-1}}} \times T^{V_{b-2} \times 2^{(b-2)L_{b-2}}} \times \cdots \times T^{V_i \times 2^{iL_i}} \times \cdots \times T^{V_1 \times 2^{L_1}} \times T^{V_0}$$

$$= \left(\left(\left(\left(T^{V_{b-1}}\right)^{2^{L_{b-1}}} \times T^{V_{b-2}}\right)^{2^{L_{b-2}}} \times \cdots \times T^{V_i}\right)^{2^{L_i}} \times \cdots \times T^{V_1}\right)^{2^{L_1}} \times T^{V_0}$$

wherein the computation of line 4 yields the first term, that of line 6 performs the exponentiation and finally in that of line 7 is a conditional multiplication, i.e. if the current partition P_i contains only zero-digits, then $V_i = 0$ and therefore the computation of line 7 will keep C unchanged. Observe when E^* is used, the value V_i of a partition P_i can be null, positive or negative. When V_i is negative the multiplication of line 8 in the algorithm of Fig. 1, is actually a division.

Algorithm (*Recoding*)*ModularExpo*(T, M, E)

1: Partition E^*/E using the given strategy;
2: **for** each P_i in $\wp\,(E^*|E)$
3: Compute T^{V_i} mod M;
4: $C = T^{V_{b-1}}$ mod M;
5: **for** i = b-2 **downto** 0
6: $C = T^{2^{L_i}}$ mod M;
7: **if** $V_i \neq 0$ **then** $C = C \times T^{V_i}$ mod M;
8: **end**;
9: **return** C;
End.

Fig 1: (Recoding) Modular exponentiation algorithm

The algorithm based on a given addition-subtraction chain, used to compute the modular exponentiation $C = T^E$ mod M, is specified by the algorithm of Fig. 2.

Algorithm additionSubtractionChainBasedMethod(T, M, E)

1: Let ($a_0=1$ a_1 a_2 ... $a_l=E$) be an addition-subtraction chain for E;
2: powerOfT[0] = T mod M;
3: **for** k = 1 **to** l
4: **let** $a_k = a_i \pm a_j$ | i<k and j<k;
5: **if** ($a_k == a_i + a_j$) **then**
6: powerOfT[k] = powerOfT[i]×powerOfT[j] mod M;
7: **else**
8: powerOfT[k] = powerOfT[i]×(powerOfT[j])^1 mod M;
9: **end**;
9: **return** powerOfT[l];
End.

Fig. 2: Addition-subtraction chain based methods algorithm

A *redundant* addition-subtraction chain is an addition-subtraction chain in which some of its members are repeated otherwise, it is said *non-redundant*. Let *nrasc* (*non-r*edundant *a*ddition-*s*ubtraction *c*hain) be a function whose application to a redundant addition-subtraction chain yields the corresponding non-redundant addition-subtraction chain. For instance, $nrasc((1, 2, 2, 7, 5, 10)) = (1, 2, 7, 5, 10)$. Assuming that there are b partitions in E, then the (recoding) m-ary and the (recoding) window methods computes $C = T^E \bmod M$ using the following addition-subtraction chain (for details see [11]):

$$nrasc \left(\begin{array}{l} V_{b-1}, \left(V_{b-1}\right)^{2^{L_{b-2}}}, \left(V_{b-1}\right)^{2^{L_{b-2}}} + V_{b-2} \times \delta_{b-2}, \quad \cdots, \\ \left(\left(\left(V_{b-1}\right)^{2^{L_{b-2}}} + V_{b-2} \times \delta_{b-2} \right)^{2^{L_{b-3}}} + V_{b-3} \times \delta_{b-3} + \ldots + V_1 \times \delta_1 \right)^{2^{L_0}} + V_0 \times \delta_0 \end{array} \right)$$

where δ_i is 0 whenever P_i is a zero partition (i.e. 00...0) and 1 otherwise (i.e. contains at least one digit different from 0). Note that we apply function $nrac$ because when δ_i is 0 the current member is the same as the previous one of the chain. Note that V_i can be negative when the recoded exponent is used.

With the (recoding) m-ary method, partition length is constant and so the addition-subtraction chain simplifies as follows:

$$nrasc \left(\begin{array}{l} V_{b-1}, \left(V_{b-1}\right)^{2^L}, \left(V_{b-1}\right)^{2^L} + V_{b-2} \times \delta_{b-2}, \ldots, \\ \left(\left(\left(V_{b-1}\right)^{2^L} + V_{b-2} \times \delta_{b-2} \right)^{2^L} + V_{b-3} \times \delta_{b-3} + \ldots + V_1 \times \delta_1 \right)^{2^L} + V_0 \times \delta_0 \end{array} \right)$$

In order to guarantee that given lists above are valid addition-subtraction chains, we need to show how a term V^{2^L} is decomposed: $(V, V^2, V^4, \ldots, V^{2^L})$.

The addition-subtraction chains given above are still incomplete because they do not include the initial pre-computing addition chain. The m-ary pre-computing addition-subtraction chain is $(1, 2, 3, \ldots, m-1)$ while for that of the sliding windows is $(1, 2, 3, 5, \ldots, 2^d-1)$ wherein d is the length of the longest partition of the exponent.

Particularly, in the 1-ary method or simply the binary method, each digit represents a partition. So we can safely write $V_i = e_i$, where e_i is the ith binary digit of exponent E. Hence the addition-subtraction chain used by the (recoding) binary method is as follows:

$$nrac \left(e_{b-1}, \left(e_{b-1}\right)^2, \left(e_{b-1}\right)^2 + e_{b-2}, \ldots, \left(\left(\left(e_{b-1}\right)^2 + e_{b-2} \right)^2 + e_{b-3} + \ldots + e_1 \right)^2 + e_0 \right)$$

wherein constant b is the number of bits in the binary representation of E. Here, we do not need to use δ_i as it plays the same role of e_i. In the case of the recoding binary method e_i^*, which the ith digit of the recoded exponent E^* should substitute e_i in the chain (3). Recall that e_i^* can be $-1(\bar{1})$, 0 or 1.

A variation of the sliding window method can be found in [14], [15]. For instance, if $E = 55_{10} = 110111 = \underline{100}1\underline{001}$, the non-redundant addition chain used depending on m is as in Table 1 where the underlined part indicated the necessary pre-computing. In the case of (recoding) window method, the underline in the exponent representation indicates the partitioning used. The two last columns of Table 1 indicate the number of multiplications necessary as well as the number of inversing operations.

Table 1: Comparison of the Addition-Subtraction Chain for the Exposed Methods E = 55

Method	Addition-subtraction chain	# of Mult	# of inv
Binary	(1, 2, 3, 6, 12, 13, 26, 27, 54, 55)	9	0
Recoding binary	(1, 2, 4, 8, 7, 14, 28, 56, 55)	8	2
Quaternary	(1, 2, 3, 6, 12, 13, 26, 52, 55)	8	0
Recoding quaternary	(1, 2, 3, 4, 8, 16, 14, 28, 56, 55)	10	2
Octal	(1, 2, 3, 4, 5, 6, 7, 12, 24, 48, 55)	10	0
Recoding octal	(1, 2, 3, 4, 5, 6, 7, 8, 14, 28, 56, 55)	11	1
Window	(1, 2, 3, 4, 5, 6, 7, 12, 24, 48, 55)	10	0
Recoding window	(1, 2, 3, 4, 8, 7, 14, 28, 56, 55)	9	2

The complexity of the (recoding) m-ary methods and (recoding) window methods are scattered over the literature. For the m-ary methods, it has been shown that in general it requires $2^r-2+k-r+(k/r-1)(1-2^{-r})$ multiplications on the average, wherein k is the total number of digits in the binary representation of E and $r = 2^m$ is the number of digits in each partition. Compared with the binary method, the m-ary method for $m \neq 2$ should at most yield an asymptotic saving of 33% in terms of required multiplications.

In the case of the sliding window methods, the complexity was studied in [13]. An approximation of the lower bound of the average number of multiplications required, say A, can be computed as follows, wherein k is, as before, the total number of digits in the binary representation of E and r is the maximum number of digits allowed in each partition.

$$A \approx k - \left(r - \frac{p - p^k}{1 - p} \right) + \frac{k}{r + \dfrac{p}{1 - p}} + 2^{r-1}$$

The average number of multiplications can be found by modelling the partitioning process as a Markov chain [12]. Compared with the m-ary methods (for m chosen adequately to yield the best performance possible) for a given exponent of large size, the sliding window methods with constant-length partitions require 37% less multiplications while those that allow variable-length windows require 58% fewer multiplications [11].

It has been established in [5] that the average number of multiplications required by the recoding binary method for large exponents can be reduced by a factor of $\frac{3}{4}k+O(1)$, compared with the binary method. The asymptotical savings introduced reaches 11%.

3. PRINCIPLES OF GENETIC ALGORITHMS

Genetic algorithms maintain a *population* of *individuals* that evolve according to *selection* rules and other *genetic operators*, such as *mutation* and *recombination*. Each individual receives a measure of *fitness*. *Selection* focuses on high fitness individuals. Mutation and recombination provide general heuristics that simulate the reproduction or *crossover* process. Those operators attempt to perturb the characteristics of the parent individuals as to generate *distinct* offspring individuals.

Genetic algorithms are implemented through the following algorithm described by the algorithm of Fig. 3, wherein parameters *populationSize*, *fit* and *generationNumber* are the population maximum size, the expected fitness of the returned individual and the maximum number of generation allowed respectively. The parameter *fit* represents the length of the addition-subtraction chain that the genetic algorithm should return. Let *length(E)* be the length of the minimal addition-subtraction chain for exponent E. The exact value of *length(E)* is known only for relatively small values of E. Nevertheless, for addition chains, it is known that for large exponents E,

$$length(E) = \log E + (1 + O(1)) \frac{\log E}{\log \log E}$$

The lower bound was shown by Erdös [6] using a counting argument and the upper bound is given the *m*-ary method. So the parameter *fit* can be approximated using equation 4.

In the algorithm of Fig. 3, Function *intialPopulation* returns a valid random set of individuals that would compose the population of first generation, function *evaluate* returns the fitness of a given population storing the result into *fitness*. Function *select* chooses according to some random criterion that privilege fitter individuals, the individuals that should be used to generate the population of the next generation and function *reproduction* implements the crossover and the mutation process to actually yield the new population.

Algorithm GA(populationSize,fit,generationNumber)
1:   generation = 0;
2:   population = initialPopulation();
3:   fitness = evaluate(population);
4:   **do**
5:   parents = select(population);
6:   population= reproduce(parents);
7:   fitness = evaluate(population);
8:   generation= generation + 1;
9:   **while**(fitness[i]<fit, $\forall$ i∈population)&(generation<generationNumber);
   10: **return** fittestIndividual(population);
End.

Fig. 3: Genetic algorithms basic cycle

4. APPLICATION TO ADDITION-SUBTRACTION CHAIN MINIMISATION PROBLEM

It is perfectly clear that the shorter the addition chain is, the faster the algorithm of Fig 2. We propose a novel idea based on genetic algorithm to solve the addition chain minimisation problem. It consists of finding a sequence of numbers that constitutes an addition chain for a given exponent. The sequence of numbers should be of a minimal length.

4.1. Individual Encoding

Encoding of individuals is one of the implementation decisions one has to take in order to use genetic algorithms. It depends on the nature of the problem to solve. There are several representations that have been used with success: *binary encoding* which is the most common mainly because it was used in the first works on genetic algorithms, represents an individual as a string of bits; *permutation encoding* mainly used in ordering problem, encodes an individual as a sequence of integer; *value encoding* represents an individual as a sequence of values that are some evaluation of some aspect of the problem [16], [18].

In our implementation, an individual represents an addition-subtraction chain. We use the binary encoding wherein 1 implies that the entry number is a member of the addition-subtraction chain and 0 otherwise. Observe that an individual representing exponent E has $E+1$ entries. Let $n = 11$ be the exponent, the encoding of Fig. 4 represents the addition chain $(1, 2, 3, 6, 12, 11)$:

1	2	3	4	5	6	7	8	9	10	11	12
1	1	1	0	0	1	0	0	0	0	1	1

Fig. 4: Addition-subtraction chain encoding

This encoding allows invalid solutions to be represented. Obviously, this increases the search space. On the other hand, a more sophisticated encoding (using a matrix or tree-based encoding), with which only valid chains can be represented, complicates dramatically the genetic algorithm and so degrades its overall performance. There are two different ways to avoid the reproduction of individuals that are not a potential solution (i.e. not an addition-subtraction chain): *(i)* Every time a new individual is yield, it is checked whether it represents an addition-subtraction chain and if so discarded and another offspring is generated; *(ii)* The invalid solutions can efficiently swept away as soon as they are introduced into the population, if these are declared exaggeratedly unfit using the penalty method. This is used and will be explained later on.

4.2. The Genetic Algorithm

Consider the genetic algorithm of Fig. 3. Besides the parameters *populationSize*, *fit* and *generationNumber* which represent the population maximum size, the fitness of the expected result and the maximum number of generation allowed, the genetic algorithm has several other parameters, which can be adjust by the user so that the result is up to his or her expectation. The selection is performed using some *selection probabilities* and the reproduction, as it is subdivided into crossover and mutation processes, depends on the kind of crossover and the mutation rate and degree to be used.

4.2.1 Selection

The selection problem consists of how to select the individuals that should yield the new population. According to Darwin's evolution theory the best ones should survive and create new offspring. There are many selection methods [1], [7]. These methods include *roulette wheel* selection or *fitness proportionate reproduction* and *rank* selection. In the following, we describe the idea behind each of these selection methods. In our implementation, we use fitness proportionate reproduction.

Fitness Proportionate Reproduction.

Parents are selected according to their fitness. The better the fitness the individuals have, the higher their chances to be selected are. Imagine a roulette wheel where are placed all individuals of the population, wherein every individual has portion proportionate to its fitness, as it is shown in Fig 5.

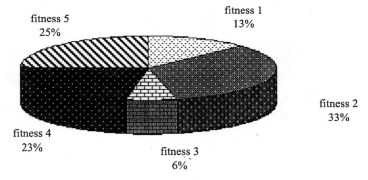

Fig. 5: Representation of the roulette wheel selection

Then a marble is thrown into the roulette and selects an individual. It is clear that individuals with bigger portion in the wheel will be selected more times. The selection process can be simulated by following steps: *(i)* first sum up the fitness of all individuals in the population and let S be the obtained sum; *(ii)* then generate a random number from the [0, S], and let f be this number; *(iii)* go through the individuals of the population, summing up the fitness of the next one. Let σ be this partial sum; *(iv)* If $\sigma \geq f$, then stop the selection process and choose the current individual otherwise return to step *(ii)*.

Rank Selection

The fitness proportionate reproduction selection presents some limitations when the individual fitnesses differ too much from one another. For instance, if the best individual has a fitness of 95% of the entire roulette wheel then the other individuals will have very few, if any, chances to be selected. To get round this limitation, the *rank selection* method first ranks the individuals of the population according to its corresponding fitness. The individual with the worst fitness receives rank1 and that with the best fitness receives rank N, which is the number of individuals in the population. The impact of the ranking process is shown in Fig 6, which represents the roulette wheel before and after the ranking of the individuals. Rank selction may yield a slower convergence as the fittest individuals and those that are less fit have much closer fitnesses.

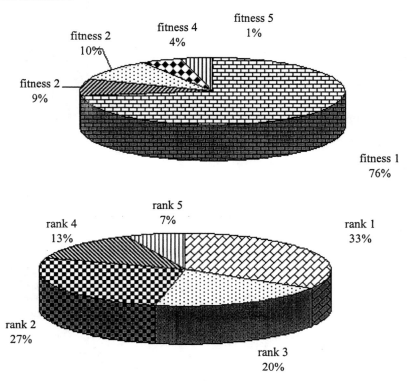

Fig. 6: Representation of the roulette wheel selection before and after ranking the individuals according to their fitness.

4.2.2 Reproduction

Given the parents populations, the reproduction can proceed using different schemes [1], [7]: a *total* replacement, *steady-state* replacement and *elitism*. In the first scheme, offspring replace their parents in the population of the next generation. That is only offspring are used to form the population of the next generation. The steady-state replacement exploits the idea that only few low-fitness individuals should be discarded in the next generation and should then be replaced by offspring. Finally, elitism exploits the idea that the best solution might be the fittest individual of the current population and so transports it unchanged into the population of the next generation. In our implementation we use the total replacement reproduction scheme as well as elitism.

Obtaining offspring that share some traits with their corresponding parents is performed by the *crossover* function. There are several *types* of crossover operators. These will be presented shortly. The newly obtained population can then suffer some mutation, i.e. some of the individuals of some of the genes. The crossover type, the number of individuals that should mutated and how far these individuals should be altered are set up during the initialisation process of the genetic algorithm.

Crossover

There are many ways on how to perform crossover and these may depend on the individual encoding used [12]. We present crossover techniques used with permutation representation. *Single-point crossover* consists of choosing randomly one *crossover point* then, the part of the integer sequence from beginning of offspring till the crossover point is copied from one parent, the rest is copied from the second parent as depicted in Fig. 7(a). *Double-points crossover* consists of selecting randomly two *crossover points*, the part of the integer sequence from beginning of offspring to the first crossover point is copied from one parent, the part from the first to the second crossover point is copied from the second parent and the rest is copied from the first parent as depicted in Fig. 7(b). *Uniform crossover* copies integers randomly from the first or from the second parent. Finally, *arithmetic crossover* consists of applying some arithmetic operation to yield a new offspring.

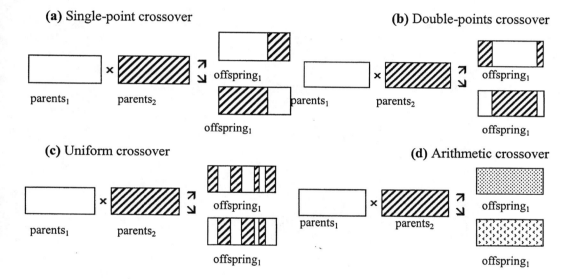

Fig. 7: Different types of crossover

The single point and two points crossover may use randomly selected crossover points to allow variation in the generated offspring and to contribute in the avoidance of premature convergence on a local optimum [2]. In our implementation, we tested all four-crossover strategies. We implemented both single and double-point crossover. We found out that with the latter the evolution is more efficient (see the implementation results in the next section).

Mutation

Mutation consists of altering some genes of some individuals of the population obtained after crossover. The number of individuals that should be mutated is given by the parameter *mutation rate* while the parameter *mutation degree* states how many genes of a selected individual should be changed. The mutation parameters have to be chosen carefully as if mutation occurs very often then the genetic algorithm would in fact change to *random search* [2]. When either of the mutation rate or mutation degree is null, the population is then kept unchanged, i.e. the population obtained from the crossover procedure represents actually the next generation population.

The essence of the mutation process depends on the encoding type used. When binary encoding is used, the mutation is nothing but a bit inversion of those bit genes that were randomised. When permutation encoding is used, the mutation is reduced to a permutation of some randomly selected integer genes. When value encoding is used, then a very small value is added or subtracted from the randomised genes.

As for our implementation, when mutation takes place, a number of genes are randomised and its value inverted, i.e. if the selected gene is 1 then it becomes 0 and vice-versa.

4.2.3 Fitness

This step of the genetic algorithm allows us to classify the population so that fitter individuals are selected more often to contribute in the formation of a new population. As we said before we use roulette wheel selection

The fitness evaluation of an addition-subtraction chain is done with respect to two aspects: *(i)* how much a given addition-subtraction chain adheres to the Definition 1, i.e. how many members of the chain cannot obtained summing up or subtracting from one another two previous members of the chain; *(ii)* how far the addition chain is reduced, i.e. what is the length of the addition-subtraction chain. The algorithm of Fig. 8 describes the evaluation procedure of fitness used in our genetic algorithm.

Algorithm evaluate(individual a)
1:   **int** fitness = 0;
2:   **for** i = 1 **to** exponent **do**
3:   **if** a[i] == 1 **then**
4:   fitness = fitness + 1;
5:   **if** ∃ j,k| 1≤k≤j≤i **and** (i==j+k **or** i==j-k) **and** a[i]==a[k]==1 **then**
6:   fitness = fitness + largePenalty;
7:   **end**;
8:   **return** fitness;
End.

Fig. 8: Algorithm of the evaluation function

The use of penalty in the fitness function is needed because the encoding allows invalid solution to be represented. For valid individual, the fitness function represents the length of the addition-subtraction chain represented. The evolutionary process attempts to minimise the number of ones in a valid addition-subtraction chain and hence minimise the corresponding length. Individuals with fitness bigger than the exponent are invalid addition-subtraction chains as the constant *largePenalty* is chosen to be larger than the exponent. With well-chosen parameters, the genetic algorithm deals only with valid addition chains.

4.2.4 Implementation Results

In applications of genetic algorithms to a practical problem, it is difficult to predict a priori what combination of settings will produce the best result for the problem. The settings consist of the population size, the crossover type, the mutation rate, when mutation does take place, the mutation degree. We investigated the impact of different values of these parameters in order to choose the more adequate ones to use. We found out that the ideal parameters are: a population of 100 individuals; the double-points crossover; a mutation rate between 0.1 and 0.3 and a mutation degree of about 1% of the value of the exponent.

The charts of Fig. 10 shows the progress made in the first 200 generations of an execution to obtain the addition chain for 250. We exploited the single-point crossover in the first evolution and the double-point crossover in the second one. The other parameter settings used are: a 100 individual by population, a mutation rate of 0.1 and a mutation degree of 3.

Finding the best addition-subtraction chain is impractical. However, we can find near-optimal ones. Our genetic algorithm always finds addition-subtraction chains as short as the shortest and sometimes shorter than those chains used by the (recoding) m-ary method independently of the value of m and by the (recoding) windows independently of the partitioning strategy used.

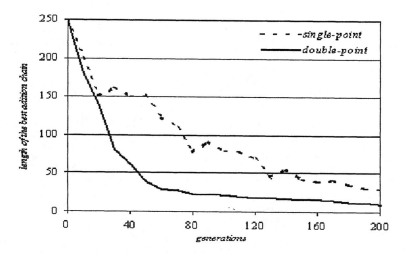

Fig. 9: Genetic algorithm result curve for the parameters given above

Here are some examples of addition chains for exponents 23, 55 and 250. The choice of these numbers is completely arbitrary. In Table 2, we compare the addition-only chains obtained by the exposed methods vs. those yield by our genetic algorithm while, we do the same thing with addition-subtraction chains. For exponent 23, 55 and 250, the use optimal addition chain requires more than 5, 6 and 8 multiplications. The genetic algorithm finds addition chains that are shorter that those obtained using both (recoding) m-ary and window methods.

The addition chains of Table 3 and For the window method, the exponent are partitioned as follows: $E = (23)_{10} = 10111_2$; $E = 55_{10} = 110111_2$ and $E = 250_{10} = 11111010_2$. The addition-subtraction chains of Table 3, the exponents were recoded as follows: $E = (23)_{10} = 10\underline{1}00\underline{1}$; $E = 55_{10} = 100\underline{1}00\underline{1}$ and $E = 250_{10} = 1000\underline{1}0\underline{1}0$. The underlines indicate the window partitioning.

The execution times of the genetic algorithm for finding the addition and addition-subtraction chains above are given in Table 4.

**Table 2: Comparison of the Addition Chain Length
for the Exposed Methods E = 23, 55, 250**

Method	Addition chain	# of Mult
Binary	(1, 2, 4, 5, 10, 11, 22, **23**)	7
	(1, 2, 4, 6, 12, 13, 26, 27, 54, **55**)	9
	(1, 2, 3, 6, 7, 14, 15, 30, 31, 62, 124, 125, **250**)	12
Quaternary	(1, 2, 3, 4, 5, 10, 20, **23**)	7
	(1, 2, 3, 6, 12, 13, 26, 52, **55**)	8
	(1, 2, 3, 6, 12, 15, 30, 60, 62, 124, 248, **250**)	11
Octal	(1, 2, 3, 4, 5, 6, 7, 8, 16, **23**)	9
	(1, 2, 3, 4, 5, 6, 7, 12, 24, 48, **55**)	9
	(1, 2, 3, 4, 5, 6, 7, 12, 24, 31, 62, 124, 248, **250**)	13
Window	(1, 2, 3, 5, 7, 8, 16, **23**)	7
	(1, 2, 3, 5, 7, 12, 24, 48, **55**)	8
	(1, 2, 3, 5, 7, 8, 15, 30, 60, 120, 125, **250**)	11
Genetic algorithm	(1, 2, 3, 5, 10, 20, **23**)	6
	(1, 2, 4, 8, 9, 18, 27, 28, **55**)	8
	(1, 2, 4, 8, 10, 20, 40, 80, 120, 240, **250**)	10

Table 3: Addition-Subtraction Chains for the Exposed Methods vs. GAs (E =23, 55, 250)

Method	Addition-subtraction chain	# of Mult	# of Div
Recoding binary	(1, 2, 4, 3, 6, 12, 24, **23**)	7	2
	(1, 2, 4, 8, 7, 14, 28, 56, **55**)	8	2
	(1, 2, 4, 8, 16, 32, 31, 62, 124, 125, **250**)	10	1
Recoding quaternary	(1, 2, 4, 8, 6, 12, 24, **23**)	7	2
	(1, 2, 3, 4, 8, 16, 14, 28, 56, **55**)	10	2
	(1, 2, 4, 8, 16, 32, 64, 62, 124, 248, **250**)	10	1
Recoding octal	(1, 2, 3, 6, 12, 24, **23**)	6	1
	(1, 2, 3, 4, 5, 6, 7, 8, 14, 28, 56, **55**)	11	1
	(1, 2, 4, 8, 16, 32, 31, 62, 124, 248, **250**)	10	1
Recoding window	(1, 2, 4, 3, 6, 12, 24, **23**)	7	2
	(1, 2, 3, 4, 8, 7, 14, 28, 56, **55**)	9	2
	(1, 2, 4, 8, 16, 32, 31, 62, 124, 125, **250**)	10	1
Genetic algorithm	(1, 2, 3, 6, 12, 11, **23**)	6	1
	(1, 2, 4, 8, 16, 24, 32, 56, **55**)	8	1
	(1, 2, 3, 4, 8, 16, 32, 64, 128, 125, **250**)	10	1

Table 4: Execution times of the genetic algorithm

Type	Chain	Execution time (ms.)
Addition chain	(1, 2, 3, 5, 10, 20, **23**)	1.89
	(1, 2, 4, 8, 9, 18, 27, 28, **55**)	2.01
	(1, 2, 4, 8, 10, 20, 40, 80, 120, 240, **250**)	7.61
Addition-subtraction chain	(1, 2, 3, 6, 12, 11, **23**)	1.17
	(1, 2, 4, 8, 16, 24, 32, 56, **55**)	1.97
	(1, 2, 3, 4, 8, 16, 32, 64, 128, 125, **250**)	6.53

For larger exponent, a comparison of the performance of the *m*-ary, sliding window vs. the genetic algorithm is shown in Fig. 12. A *good* addition-subtraction chain for exponent of 128-bits (i.e. in the binary representation) can be obtained in at most 6 seconds using a Pentium III with a 128 MB of RAM.

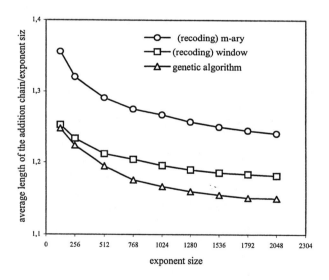

Fig. 10: Average length of the addition-subtraction chain/exponent size yield by the genetic algorithm vs. the(recoding) m-ary and (recoding) window methods.

5. CONCLUSIONS

In this paper, we presented an application of genetic algorithms to minimisation of addition-subtraction chain. We first explained how individuals are encoded. Then we described the necessary algorithmic solution. Then we presented some empirical observations about the performance of the genetic algorithm implementation.

This application of genetic algorithms to the classical minimisation proved to be very useful and effective technique. Shorter addition-subtraction chains compared with those obtained by the (recoding) m-ary methods as well as those obtained for the (recoding) window methods (see Table 2 and 3 of the previous section) can be obtained with a little computational effort. A comparison of the overall performance of the m-ary, sliding window vs. the genetic algorithm was provided.

ACKNOWLEDGMENTS

The authors would very much like to thank the referees for their critical reading of this paper and their very useful observations. They contributed very positively to the content of this paper.

REFERENCES

[1]   Davis, L., *Handbook of Genetic Algorithms*, Van Nostrand Reinhold, New York, 1991.

[2]   DeJong, K. and Spears, W.M., *An analysis of the interacting roles of the population size and crossover type in genetic algorithms*, In Parallel problem solving from nature, pp. 38-47, Springer-Verlag, 1990.

[3]   DeJong, K. and Spears, W.M., *Using genetic algorithms to solve NP-complete problems*, Proceedings of the Third International Conference on Genetic Algorithms, pp. 124-132, Morgan Kaufmann, 1989.

[4]   P. Downey, B. Leong and R. Sethi, *Computing sequences with addition chains*, SIAM Journal on Computing, vol. 10, pp. 638-646, 1981.

[5]   Eğecioğlu, Ö. and Koç, Ç.K., *Fast modular exponentiation*, Proceedings of International Conference on New Trends in Communication, Control and Signal Processing, Ankara, Turkey, pp. 188-194, 1990.

[6]   Erdös, P., *Remarks on number theory III: On addition chain*, Acta Arithmetica, pp. 77-81, 1960.

[7]   Goldberg, D. E., *Genetic Algorithms in Search, Optimisation and Machine Learning*, Addison-Wesley, Massachusetts, Reading, MA, 1989.

[8]   Gordon, D.M., *A survey of fast exponentiation methods*, technical report

[9]   Haupt, R.L. and Haupt, S.E., *Practical genetic algorithms*, John Wiley and Sons, New York, 1998.

[10]   Knuth, D.E., *The Art of Programming: Seminumerical Algorithms*, vol. 2. Reading, MA: Addison Wesley, Second edition, 1981.

[11]   Koç, Ç.K., *High-speed RSA Implementation*, Technical report, RSA Laboratories, Redwood City, California, USA, November 1994.

[12]   Koç, Ç.K., *Analysis of sling window techniques for exponentiation*, Computers and Mathematics with Applications, vol. 30, no. 10, pp. 17-24, 1995.

[13]   Kunihiro, *Study on fast algorithm for exponentiation*, M.E. thesis, Department of Mathematics, Engineering and Informatics, University of Tokyo, 1996.

[14]   Kunihiro, N. and Yamamoto, H., *New methods for generating short addition chain*, IEICE Transactions, vol. E83-A, no. 1, pp. 60-67, January 2000.

[15] Kunihiro, N. and Yamamoto, H., *Window and extended window methods for addition chain and addition-subtraction chain*, IEICE Transactions, vol. E83-A, no. 1, pp. 72-81, January 1998.

[16] Michalewics, Z., *Genetic algorithms + data structures = evolution program*, Springer-Verlag, USA, third edition, 1996.

[17] National Institute of Science and Technology – NIST, *A proposed Federal information processing standard for digital signature standard - DSS*, Federal Register 56, p. 169, 42980-42982, 1991.

[18] Neves, J., Rocha, M., Rodrigues, Biscaia, M. and Alves, J., *Adaptive strategies and the design evolutionary applications*, Proceedings of the Genetic and the Design of Evolutionary Computation Conference, Orlando, Florida, USA, 1999.

[19] O'Connor, L., *An Analysis of Exponentiation Based on Formal Languages*, in Proceeding of Advances in Cryptology – EUROCRYPT'99, International Conference on the Theory and Application of Cryptographic Techniques, Prague, Czech Republic, Lecture Notes in Computer Science 1592, Springer 1999.

[20] Rivest, R.L., Shamir, A. and Adleman, L., *A method for obtaining digital signature and public-key cryptosystems*, Communication of ACM, vol. 21, no.2, pp. 120-126, 1978.

In: Biocomputing
Editor: Phillip A. Laplante, pp. 73-89

ISBN 1-59033-889-8
2003 © Nova Science Publishers, Inc.

Chapter 6

DYNAMIC DNA COMPUTING MODEL

Zhiquan Frank Qiu and Mi Lu

Department of Electrical Engineering
Texas A&M University
College Station, Texas 77843-3128
U.S.A.
{zhiquan, mlu}@ee.tamu.edu

ABSTRACT

It has been evidenced that DNA computing can solve those problems which are currently intractable on the even fastest electronic computers. The algorithm design for DNA computing, however, is not straightforward: A strong background in both DNA molecule and computer engineering are required to develop efficient DNA computing algorithms. All of these algorithms need to start over from the very beginning when their initial condition changes. This can be frustrating, especially if the change in the initial condition is very small. The existing models from which a few DNA computing algorithms were developed are not able to accomplish this dynamic updating.

For a long time, people have talked about the huge memory made possible through DNA computing due to the fact that each strand can be treated as both storage media and processor. There is, however, no existing application that has yet made use of this huge memory because, though, it is very easy to read from this memory, it is very difficult to write data to it. The memory can only be read after the data has been stored.

In this paper, a new DNA computing model is introduced based on which new algorithms are developed to solve the 3-Coloring problem. These new algorithms are presented as vehicles for demonstrating the advantages of the new model, and can be expanded to solve other NP-complete problems. They have the advantage of dynamic updating, so an answer can be changed based on modifications to the initial condition. The new model makes use of this huge memory by generating a "lookup table" during the process of implementing the algorithms. If the initial condition changes, the answer will change accordingly. In addition, the new model has the advantage of decoding all the answer strands in the final pool very quickly and efficiently. The advantage provided by this new model makes DNA computing both efficient and attractive in solving computationally intense problems.

Key-Words: - DNA Computing, Molecular Computing, Coloring Problem, Parallel
Processing, Dynamic Updating, Efficient Decoding

1. INTRODUCTION

1.1. Motivation

A strand of DNA is composed of four different base nucleotides: A(adenine),
C(cytosine), G(guanine) and T(thymine). When attached to deoxyribose, these base
nucleotides can be strung together to form a strand. Because DNA strands can be used to
encode information and DNA bio-operations are completed based on the interactions between
strands, each DNA strand can be counted as a processor as well as storage media. Numerous
strands are involved in DNA bio-operations and the interactions between one another occur
simultaneously. This, then, can be viewed as a realization of massive parallel processing.

Since Adleman [1] solved a 7-vertex instance of the Hamiltonian Path Problem, a well
known representative of NP-complete problems, the major goal of subsequent research in the
area of DNA computing has been to develop new techniques for solving NP-complete
problems that cannot be solved by current electronic computers in a reasonable amount of
time. NP-complete problems are those problems for which no polynomial-time algorithm has
yet been discovered, in contrast to polynomial-time algorithms whose worst-case run time is
$O(n^k)$ for some constant k, where n is the size of the problem.

Consider that one liter of water can hold 10^{22} DNA strands. The potential computing
power is significant, and this recognition raises the hope of solving problems currently
intractable on electronic computers. Rather than use electronic computers upon which the
time needed to solve NP-complete problems grows exponentially with the size of the
problem, DNA computing technology can be used to solve these problems within a time
proportional to the problem size. An NP-complete problem which may take thousands of
years for current electronic computers to solve would take a few months, if the existing DNA
computing techniques were adopted.

As indicated in several articles [8][9][10][11][13][15], most DNA computing algorithms
are based on certain developed DNA computing models. The most popular models are the
sticker based model [16][17], the surface based model [14][18]. and the self-assembly based
model [20][19]. The problem with the sticker based model is that the stickers annealed to the
long strand may fall off during bio-operations, thus causing a very high rate of error. The
limitation of the surface based model is that the scale of computation is severely restricted by
the 2-dimensional nature of surface based computations. The shortcoming of the self-
assembly based model is that it makes use of biological operations not yet matured.

While the theory of molecular computation has developed rapidly, most of these
algorithms usually take months to solve problems that may then take thousands of years to
solve with electronic computers. The problem is that when the initial condition changes, the
algorithms have to start over again. In this paper, a new DNA computing model is introduced
which can eliminate this problem. Based on this model, algorithms can be designed to
dynamically update the answer. When the initial condition changes, the new algorithms can
continue with the current process, and the solution for the new problem can be generated by a

few extra processes. In addition, this new model can also be used to solve several similar problems simultaneously. Even more attractively, our new model can decode all the answer strands very quickly and cost efficiently.

1.2. Our New Model

Our new model adopts only the mature DNA biological operations [1][13]. The following basic principle operations: *synthesis, ligation, separation, combination* and *detection* are selected for building the new model.

- **synthesis** *I(P, π)* To generate a pool of coded strands, *P*, following criteria π. Strands are coded differently for different applications by using the four base nucleotides: A, G, T and C. A set is defined as a group of strands, and the container holding a set of strands is called a pool. If the criteria are the colors of a node in a graph, then a pool of strands coding all the possible colors for the node is expected after *synthesis*. In the graph coloring problem, the strand is encoded for the colors of a number of nodes. Here, a few consecutive nucleotides on the strand coded for the color of one node form a region. For example, in Figure 1, one strand consists of three regions such that *s={RBR }* where *(CCAAG)*, *(AATTC)* and *(CCAAG)* represent the colors for three nodes as *R (Red), B (Blue)* and *R (Red)*, respectively.

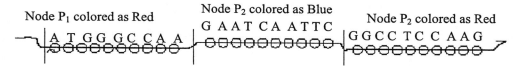

Node P_1 colored as Red

Node P_2 colored as Blue

G AAT CA ATTC

Node P_2 colored as Red

A T GG GC CA A

GGCC TC C AAG

Fig. 1: An example of three nodes in a graph that are colored by r(red), b(blue) and r(red)

- **ligation** $L(P_3, P_1, P_2)$ To bind strands in pool P_1 with strands in pool P_2. Each code s_{1_i} in P_1 is *ligated* to every other code s_{2_j} in P_2. If the strands in P_1 represent the codes $\{s_{1_i} \mid i = 1, 2, \cdots, c, where\, s_{1_i} \in P_1 \}$ and those in P_2 represent the codes $\{s_{2_j} \mid j = 1, 2, \cdots, d, where\, s_{2_j} \in P_2 \}$, after the *ligation*, the *ligated* strands are stored in P_3 and they represent the codes $\{s_k \mid k = 1, 2, \cdots, c \times d \}$ where $s_k = s_{1_i} s_{2_j}$ for $k = i + (j - 1) \times c$.

- **separation** $S(P, P_t, P_f, \theta)$ *Separation* is used to partition strands in pool P, and store those strands in two new pools: P_t and P_f based on criteria θ. After each *separation* operation, the strands that meet the criteria will be stored in one pool, P_t, while all strands that do not meet the criteria will be stored in the other pool, P_f. In order to perform the *separation* operation, many identical short strands defined as probes are attached to magnetic beads. These probes are then put into the pool containing the strands to be *separated*. Each probe can be paired up with a complementary strand to form a double helix. Such pair-ups only occur under the WC(Watson-Crick) complement rule: *A* only pairs with *T* and *G* only pairs with *C*. For example, in Figure 1, if the strands containing the region for node 1 colored as '*R*' need to be *separated*, the DNA short strands *TACCCGGTTC* should be used as a probe because *TACCCGGTTC* complements *ATGGGCCAAG*. Also, the double helix can be separated by heating in order to make the paired strands apart from each other without breaking the chemical bonds that hold the nucleotides together inside the single strand. The strands in the pool containing a region that complements the probes will be hybridized to, and captured by, the probes while all those without the region will remain in the pool [6].

A gel-based *separation* technique for DNA computing [4] has been developed which uses gel-layer probes instead of the bead to capture the strands. The capture layer only retains the strand with a region that complements the probe when it is cooled down, and will let all strands pass when the layer is heated up. The advantage of using gel-based probes over bead-based probes is that the gel-based method is more accurate when capturing DNA molecules. In Figure 2 which illustrates the gel-based *separation*, a set of strands run from the left side buffer to the right. At each capture layer, the temperature is cold in order to capture the desired strands, and all unwanted strands are passed through into one pool. Then the temperature is raised to let all desired strands in the layer pass into another pool. The strands from the left buffer are *separated* and stored in two different pools.

- **combination** *B(P, P₁, P₂)* To pour two pools, P_1 and P_2, together to form a new pool, *P*.
- **detection** *D(P)* To check if there is any strand left in the pool, *P*. If the answer is "yes", the strands in the pool should be decoded.

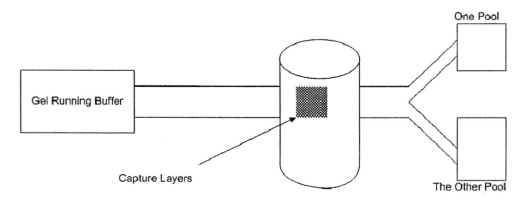

Fig. 2: separation operation based on gel layers

The rest of this paper is organized as follows: Section 2 gives an introduction to our new algorithm and how, based on the new model we proposed, it solves the 3-Coloring problem. The complexity analysis of the new algorithm is provided in section 3, which shows how dynamic updating can be accomplished. The last section concludes this paper.

2. THE NEW FUNDAMENTAL ALGORITHM

This new algorithm for the 3-Coloring problem is developed based on the new DNA computing model. The basic algorithm which will generate the answer to the 3-Coloring problem of a given graph is introduced in this section. In the next section, the algorithm will be advanced to show how the answer can be dynamically updated.

2.1. 3-Coloring Problem

The 3-Coloring problem, a special case of the k-Coloring problem where k=3, is a well known representative of the class of NP-complete problems. A new algorithm for solving the 3-Coloring problem will be introduced, and will simplify the explanation of our new DNA computing model. The algorithms developed can hereby be expanded to solve the k-Coloring problem and be generalized to solve other NP-complete problems.

k-Coloring Problem

A k-Coloring problem requires the coloring of an undirected graph $G=(V,E)$ in such a way that no two adjacent vertices share the same color [6]. The two nodes connected by an edge are referred to as adjacent vertices. The solution is a function $c: V \rightarrow 1, 2, \cdots, k$ such that $c(u) \neq c(v)$ for every edge $(u, v) \in E$. In other words, the numbers $1, 2, \cdots, k$ represent the k colors, and the adjacent vertices must have different colors. The k-Coloring problem determines whether k colors are adequate to color a given graph [7].

A simple example graph with ten nodes and ten edges, G(10,10), is given in Figure 3. It is clearly shown that the graph can be colored if $k \geq 3$.

In order to solve this 3-Coloring problem, we need to generate a pool of encoded DNA strands representing all the possible color patterns of the n-node graph where each color pattern is an assignment of colors to nodes [2][3]. For example, for nodes n_1 n_2 n_3 n_4, "*BBRG*" is one pattern which assigns Blue to n_1, Blue to n_2, Red to n_3 and Green to n_4, while "*RGBB*" is another pattern which colors n_1 n_2 n_3 n_4 as Red, Green, Blue and

Blue, respectively. After the strands are generated and stored in a pool, the strands representing the color patterns with no color conflict need to be *separated*. Two nodes along an edge are defined as having a color conflict when they share the same color. For the color patterns with some color conflicts along some edges of the graph, the corresponding strand should be filtered out from the pool.

Our new algorithm is introduced next. Following that, the dynamically updating algorithm and the advantages of the new algorithms will be described.

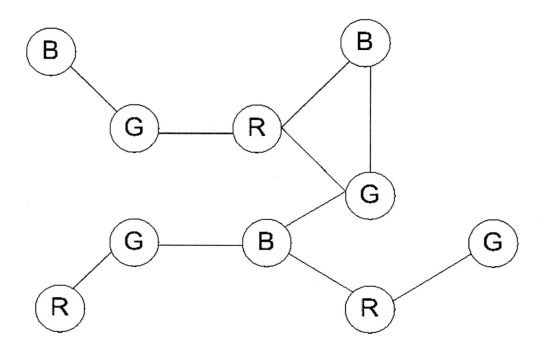

**Fig. 3. An example graph g{10,10} that can be
colored as r(red), g(green) and b(blue)**

2.2. The New Algorithm

Given a graph $G=(V, E)$, $V = \{v_i \mid i=1, 2, \cdots, n\}$ is a set of nodes and
$E = \{e_j \mid j=1, 2, \cdots, m\}$ is a set of edges. Our approach to solving the 3-Coloring problem
for such graph is divide-and-merge. Partition graph G into two subgraphs: $G_1=(V_1,E_1)$ and
$G_2=(V_2,E_2)$ such that $V_1 \bigcup V_2 = V$, $V_1 \bigcap V = \phi$ and $|V_1| \approx |V_2|$ by eliminating all edges
(u,v) such that $u \in V_1$ and $v \in V_2$. This set of edges is referred as the cut-set of G, C [6][5].
The partition process can be performed recursively. That is, subgraph G_i can be partitioned
into G_{2i+1} and G_{2i+2}, until each subgraph contains only one vertex, and n subgraphs exist in
total (See Figure 4).

After partitioning the graph G into n subgraph, the algorithm merges every two subgraphs
recursively and in parallel. Before the merge, every subgraph is colored with 3 colors. During
the merge, the color patterns of the two subgraphs are combined together. The merge
operation continues until graph G is re-established.

Note, to merge two subgraphs, the edges in the cut-sets earlier eliminated to partition the
subgraphs will be added back, and each addition of such edge will introduce a color conflict if
the nodes it links together are of the same color. Hence, the color patterns that worked for the
subgraphs may not work for the merged graph after they are combined, and some combined

color patterns will be eliminated. The elimination continues until the color patterns legitimate for the graph are found.

Our new algorithm for solving the 3-Coloring problem on a sparse graph is presented in Figure 5. The first *for* loop is used to generate n pools of strands to represent all possible color patterns for n subgraphs while initially each subgraph only contains one node.

The function of the *while* loop is, first, to merge the pairs of the two subgraphs. The bio-operation needed to merge two subgraphs is *ligation*, which *ligates* strands in two pools in order to form longer strands. Let the color patterns for subgraph G_1 be s_i and those for G_2 be s_j. For a given code s_i, all the s_j's should be *ligated* with it, and such operations are performed over all existing s_i's. That is, the strand for one color pattern of a subgraph is replicated and each duplicated copy is *ligated* with one of those strands representing the color patterns of the other subgraph. After the merge, all the color patterns of the merged graph will be represented by the *ligated* long strands.

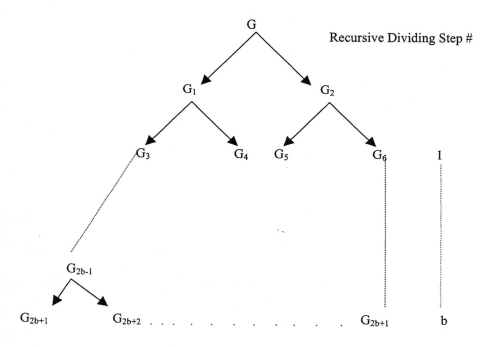

Fig. 4: Divide the graph, g_1, with $n=2^b$ nodes until each subgraph only contains one node

Inside the *while* loop, multiple copies of all the strands in all the pools need to be prepared for the next round of *ligation*. This duplication can be accomplished by using the PCR (Polymerase Chain Reaction) process [2][12].

After the merge, some *ligated* strands may encode the color patterns that have color conflicts introduced by those edges in all cut-sets eliminated in the partition step. Our task is to investigate every edge in the cut-sets and detect all the color conflicts caused hereby. This is accomplished by the *separation* operation, i.e., in all the *ligated* strands, to filter out strands that contain any color conflict from the pool. For each edge under investigation, two nodes, i and j, are connected. We first *separate* the pool into three pools that contain the strands having node i colored as R, G and B. In these three pools, the strands having node j colored as R, G and B are, respectively, filtered out by using the *separation* operation.

If there is any strand left in the final pool, P_t, then the 3-Coloring problem has an answer of "yes". Otherwise, the graph cannot be colored by only three colors and the answer is "no".

```
Algorithm 1.
for i=1 to n do
     In Parallel( I(Pᵢ, color of node i))
end
f = n
while f ≠ 1 do
     In Parallel(Make multiple copies of strands in all pools)
     for All odd i do
          In Parallel( L(Pᵢ, Pᵢ, Pᵢ₊₁) )

          In Parallel( relabel all pools 1 to  f/2 )


          for i =1 to  f/2  do
               In Parallel(
               for j = 1 to Eᵢ, Eᵢ is the number of edges in Cᵢ do
                    In Sequential { S(Pᵢ, Pᵢₜ, Pᵢ𝒻, θᵢⱼ ) },
                    θᵢⱼ is the color conflicts along eⱼ, ∀eⱼ ∈ Cᵢ
               end
               )
          end
     end

     f =  f/2

end
Check if the pool is empty to conclude "yes" or "no" accordingly.
```

**Fig. 5: The new DNA computing algorithm to
solve the 3-coloring problem for sparse graphs**

3. DYNAMICALLY UPDATING THE ANSWERS

Once a solution to the 3-Coloring problem of the graph is obtained, given minor changes in the initial conditions, it is significant to have a method that can quickly update the solution without restarting the algorithm and completely recalculating. The following is an effort toward making such dynamically updating solution both realistic and efficient.

For 3-Coloring problems, four possible changes may occur with the initial condition: nodes and edges can both be either inserted or removed. Based on the originally generated "yes" or "no" answer to the original graph, different strategies need to be considered to update the answer.

Beginning with the easiest updating strategies, if the original answer is "yes" and an edge or node is removed from the original graph, the answer will remain "yes."

If the original answer is "no", it will remain "no" when the nodes or edges are added in.

If the original answer is "yes," it may be changed to "no" after a node is inserted into the graph. An example is shown in Figure 6. In this figure, the answer to the 3-Coloring problem of the graph given in Figure 6(a) is "yes," but it changes to "no" once a node is inserted to form the graph shown in Figure 6(b).

If the original answer is "no," it may be changed to "yes" after a node is removed from the graph. Figure 7 illustrates an example. Figure 7(a) contains the graph with the answer "no" for the 3-Coloring problem. The answer changes to "yes" after one node is removed from the graph, as shown in Figure 7(b).

Inserting an edge or removing an edge can be similarly dealt with because at least one edge should be eliminated if a node is removed, and at least one edge should be added if a node is inserted.

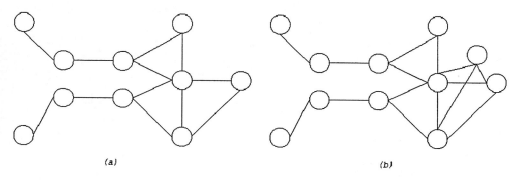

Fig. 6: Adding One Node To The Graph

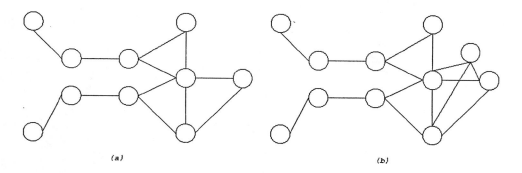

Fig. 7: Removing one node from the graph

The following illustrates how to dynamically update a solution when a node or edge is inserted into the graph, following an original answer of "yes". The strands in the final set, P_t, are checked for possible new answers. The final set is the only set that can be used because it is the only set that contains the strands that represent all of the possible coloring solutions that do not have any color conflicts among all the nodes except the newly added one. Based on these sets, only the color conflicts that occur between the newly added node and the nodes connected with it need to be checked. In other words, only the newly added edges are checked for color conflicts.

The most difficult case occurs when a node or edge is removed from a graph with an original answer of "no". The answer to the new graph may be either "yes" or "no". Removing a node includes removing both the node itself and all the edges connecting it to the graph. Following is the dynamically updating algorithm for this case. The DNA computing result that reflects an original answer of "no" is represented by an empty P_t set with no strand. All other sets represent the coloring patterns of the original graph with the color conflicts. After removing the nodes or edges, some coloring patterns may no longer have conflicts. The task then is to identify those patterns represented by DNA strands. Note that the strand sets to be examined are limited. Only those strands representing color patterns with color conflicts involving the pair of nodes connected by the edges being removed are checked. Finding the above strand sets takes $O(\alpha)$ steps, where α is the number of edges being removed. This process, in time, is much less expensive than re-computing the updated graph from the very beginning when α is not large.

The detailed algorithm needed to find the answer for the new graph with the removed edges, based on the original "no" answer, is illustrated in Figure 8.

When only one edge is removed from the original graph, pool P_{f_1} is checked. This is because P_{f_1} contains all of the strands representing all of the color combinations for the graph that have no color conflicts along all edges, except one. Assuming that the two nodes along the edge being removed are n_1 and n_2, the strands that need to be *separated* from the pool are those that have the two nodes colored as $\{RR\}$, $\{BB\}$ and $\{GG\}$. That means that only those strands which have two identically colored nodes are extracted to a new pool, P_{new}. If P_{new} is not empty, the answer to the 3-Coloring problem for the new graph is "yes", which is different from the original graph. Otherwise, the "no" answer remains.

```
Algorithm 2.
for i=1 to α do
      In Parallel( S(P_fi, P_newi, P_fi, θ_i), θ_i is the color conflicts along
                i exact # of edges)
end
B(P_new, Φ, Φ)
for i=1 to α do
      In Parallel (B(P_new, P_new, P_newi))
End
B(P_new, P_t, P_new))
for j=1 to β do
      S(P_new, P_new, P_newf, w_j), w_j is the color conflicts based on edge e_j
End
Check if the pool is empty to return the "yes" or "no" answer
accordingly.
```

Fig. 8: The dynamically updating algorithm for 3-coloring problem when α edges are removed and β edges are added

When two edges are removed from the graph, both P_{f_1} and P_{f_2} needs to be checked. This is because P_{f_2} may contain strands that represent color combinations that have color conflicts along both the edges being removed. P_{f_1} may contain strands that represent color combinations of the graph with a color conflict along only one of the two edges being removed. Suppose the two edges being removed are e_1 and e_2. Then, strands that need to be extracted from pool P_{f_2} using the *separation* operation must represent the color combinations of the graph having color conflicts along **both** edges. Strands that should be extracted from P_{f_1} are those representing color combinations with color conflict along **either** e_1 or e_2. The extracted strands are stored in a new pool, P_{new}. If P_{new} is not empty, the answer to the 3-Coloring problem for the new graph is "yes", which is different from the original graph. Otherwise, the answer for the 3-Coloring problem to the new graph remains "no".

When α different edges are removed from the original graph, α different pools should be checked. These pools are $P_{f_1}, P_{f_2}, \cdots, P_{f_\alpha}$. For different pools, different operations need to be undertaken. For pool P_{f_1}, all strands are left in due to the color conflict along one edge. If the edge that caused the conflict is removed, the answer will change to "yes". Because of this, all strands in this pool representing those color combinations with color conflicts along one of the α edges that have been removed should represent the answers to the 3-Coloring problem of the new graph. For pool P_{f_2}, all strands representing the color combinations having color conflicts along two, and only two, of the edges being removed represent the answers to the 3-Coloring problem of the new graph. For pool P_{f_t} where $t \le \alpha$, all strands representing the color combinations having color conflicts along exactly t different edges being removed will generate the answer to the 3-Coloring problem for the new graph. All strands extracted from these sets are stored in a new pool, P_{new}. If P_{new} is not empty, the answer of the 3-Coloring problem for the new graph is "yes", and thus different from the original graph. The answer is "no" if P_{new} is empty.

When the graph is changed by both removing and adding edges, multiple processing steps need to be considered. Assuming that the number of edges being removed is α and the number of edges being added is β, the strands with color conflicts along the removed edges should be found first. This will put the strands that are to be considered for the following operations in one pool, P_{new}, instead of involving several pools. Those α edges should first be considered by using the method introduced above to go through α different pools. Then, P_t is *combined* with P_{new} and relabeled P_{new}. This is due to the fact that those strands that may generate the "yes" answer are distributed in $\alpha+1$ different pools. Collecting the strands in one pool will save time and further operations as compared to working on these pools one at a time. If no strand is left in pool P_{new}, then the answer to the new graph is "no". If there are some strands in pool P_{new} after α edges are removed, color conflicts along β edges are checked. This operation can be accomplished in a manner similar to what has been described above for adding edges.

Compared to the existing algorithms, our new method can dynamically update the solution when the initial condition changes for the 3-Coloring problem of a graph. It can also solve the 3-Coloring problem for many graphs that are similar to each other. The complexity of the existing algorithms is $O(m+n)$, where n is the number of vertices and m is the number of edges [3]. If the updating process is not used, any change in the initial condition will result in a restarting of the process. With our new algorithm, the number of extra processes that need to be undertaken depends upon the significance of the changes. The complexity of the updating process is $O(\alpha + \beta)$, where α is the number of edges being removed. β is the number of edges being added.

When this method is used to solve the 3-Coloring problem for multiple graphs that are similar to each other, the time complexity is ζ after the solution for one graph is generated, where ζ is the difference between the number of edges of the two graphs.

It is necessary to check the extra space and effort that may be necessary for making dynamic updating available. First, m additional containers are needed to keep all m extra sets of strands. Second, the extra DNA material for generating these sets needs to be contained. Because strands are generated to represent all color combinations for the graph before the *separation* process takes place, no extra material is necessary as compared with the existing algorithms until the answer is generated for the original graph. The extra material is only necessary if new solutions need to be formed for the modified graph when the edges and/or nodes are added.

When the procedure for approaching a 3-Coloring problem of a given graph is finished and a new graph is provided, how can one determine whether to start again from the beginning or to use the dynamic updating method to generate the new answer?

Assume that the implementation of the algorithms introduced above for the 3-Coloring problem of the graph with n nodes and m edges has been finished, and the 3-Coloring problem of a new graph needs to be solved. This new graph has N nodes and M edges. This graph can be converted from the existing graph by first removing δ nodes and α edges, and then adding γ nodes and β edges. The new graph can be generated by changing the original graph, or it can be treated as a totally new graph. In order to solve the problem for the new graph, N *ligation* and M *separation* operations are necessary if the algorithm is being restarted from the beginning. The total time necessary is:

$$T_1 = N \times l + M \times s$$

where l is the time for each *ligation* operation and s is the time necessary for each *separation* operation. Here, *combination* operations are ignored due to their simplicity because the time needed for the *combination* operations is very short, as compared to the other operations used in DNA computing. When the answer is generated based on the pools already generated using this new, dynamically updating strategy, the time necessary for reaching the answer is:

$$T_2 = (\alpha + \beta) \times s + \gamma \times l$$

In order to take advantage of the new method, the time that is needed must be shorter than restarting the algorithm from the beginning.

$$T_2 \le T_1$$

$$(\alpha + \beta) \times s + \gamma \times l \leq N \times l + M \times s$$

$$(\alpha + \beta) \times s + \gamma \times l \leq (n + \gamma - \delta) \times l + (m + \beta - \alpha) \times s$$

because $N = n + \gamma - \delta$ and $M = m + \beta - \alpha$ It is easy to get

$$(m - \alpha) \times s + (n - \delta) \times l \geq \alpha \times s$$

As $n - \delta$ is always greater than 0, the above condition can be tightly restrained as follows:

$$(m - \alpha) \times s \geq \alpha \times s$$

So, $\alpha < \dfrac{m}{2}$. The algorithm needs to be restarted from the beginning only when the change is significant-- in other words, when more than half of the edges need to be removed to generate a new graph from the original.

Given the above conclusion, it is clear that there is no need to retain all m sets. At least half of the pools can be destroyed in order to save storage space. This will save the expenses once required for storing m sets of strands and the material needed to work on them.

4. LOCATE THE EXACT SOLUTIONS

After the final set that contains all the solutions for the 3-Coloring problem for the graph is generated, it is time to decode the strands in order to reach the color patterns that can color the graph. Each of the strands in the pool has one answer encoded and some strands in the pool may encode the same answer.

The new method introduced here can decode all the color patterns represented by the DNA strands in pool P_t without using the electron microscope to "read" the strands one by one. It is much more cost and time efficient comparing with the method that decodes the strands in the pool one at a time using electron microscope. The flow diagram of the new method is illustrated in Figure 9. The function of each box in this figure is a filter based on the gel-based *separation*. The detailed structure of each box is shown in Figure 10. The filter function is given below: before the input buffer is filled, the capture layer is filled with small segments of DNA strands. Each filter is named as *Fkc*, where $k \in \{1, 2, \cdots, n\}$ and $c \in \{R, G, B\}$, the capture layers contains the DNA strand segments that represent the color pattern complements to color c for node k. The temperature is cooled down first. Then, the input buffer lets the input pool flow into the capture layer and valve A is opened. All strands that contain the segment representing color c for node k are captured in the layer. The rest of the strands in the input pool will pass the layer and go through valve A. When this process is finished, valve B is opened and the temperature of the container is increased. All strands containing the segment that represents node k being colored with color c is *separated* from the rest of the pool. The order of the operations are indicated in Figure 9. For example, *F1r* will divide the input pool into two pools where they contain strands representing color

patterns $N_n N_{n-1} \cdots N_2 N_1 = \{XX \cdots XR\}$ and $N_n N_{n-1} \cdots N_2 N_1 = \{XX \cdots X\overline{R}\}$ where $X \in \{R, G, B\}$, $\overline{R} \in \{G, B\}$. If two filters *F1R*, and *F2R* are connected serially, it is easy to *separate* out the strands with patterns $N_n N_{n-1} \cdots N_2 N_1 = \{XX \cdots RR\}$. If n filters *Fir* where $i \in \{1, 2, \cdots, n\}$ are connected serially as shown at the left column in Figure 9, at the time strands come out from *Fnr*, these strands, if any, must be representing patterns $N_n N_{n-1} \cdots N_2 N_1 = \{RR \cdots RR\}$. If no strand comes out from this filter, then the pattern $N_n N_{n-1} \cdots N_2 N_1 = \{RR \cdots RR\}$ is not a color pattern that can color the given graph. At the time t_i, all valves labeled i in Figure 9 are opened and the temperature of the corresponding container should have been cooled down or warmed up. Eventually, x, y and z will provide some output sets. At time t_{n+1}, x should output a set. This set only has strands representing color combination of $N_n N_{n-1} \cdots N_1 = \{RR \cdots R\}$ for n nodes. The following time $t_{n+2}, t_{n+3}, \cdots, t_{n+3^n-2}$, other sets containing strands representing the color combinations $N_n N_{n-1} \cdots N_1 = \{RR \cdots RG, RR \cdots RB, RR \cdots GR, RR \cdots GB, \cdots, RB \cdots BB\}$ are outputted from x. The color combinations represented by strands are outputted from y and z in the following   order:
$$N_n N_{n-1} \cdots N_1 = \{GR \cdots RR, GR \cdots RG, GR \cdots RB, GR \cdots GR, GB \cdots GG, \cdots, GB \cdots BB\}$$
and
$$N_n N_{n-1} \cdots N_1 = \{BR \cdots RR, BR \cdots RG, BR \cdots RB, BR \cdots GR, BR \cdots GG, \cdots, BB \cdots BB\}.$$

The *decoding* process has been simplified by using the *separate* and *detect* operations. At the time a set is outputted from x, y or z, the *detect* operation will check if it is empty. If not, the corresponding color combination is a good one for coloring the graph with no color conflict along any edge. Otherwise, the set is empty and the corresponding color combination can not be used to color the graph.

Now, let's check the extra space and effort that are necessary for efficiently decoding the strands in the final set. At the beginning, it seems that $3n$ different filters are needed. When the algorithm for generating the final set is implemented, it can be clearly seen that all the filters are already generated in order to *separate* the initial pool containing strands representing all color combinations. The extra effort is needed to reorder these filters. After the filters are connected together, the valves and temperature of the containers can be controlled by electronic microcontroller automatically. The automation will greatly reduce the involvement of human beings and it will make DNA computing more error resistant. In addition, all filters on the far right column in Figure 9 are not needed because all strands coming into these strands will pass through the filter together. There is no filter function necessary here. Storage buffers can be used to replace these filters for temporarily storage in order to simplify the system. The other additional effort that is needed is the *detect* operation. This step can be accomplished very effectively and quickly.

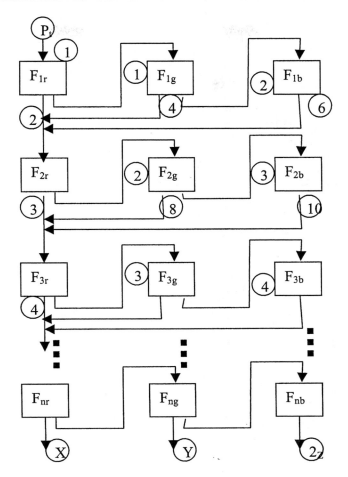

Circled numbers represent places where valves are placed

Fig. 9: Automated "decoding" process with 3n filters

5. CONCLUSION

In this paper, a new model for DNA computing is introduced. Based on the new model, our new algorithms for the 3-Coloring problem have been presented. The new algorithms have the advantage of dynamic updating, as compared to the existing algorithms. These new algorithms represent a huge improvement over the existing algorithms.

Instead of re-starting the DNA computing algorithm from the very beginning every time the initial condition would change, this new method can generate the new solution through a few extra DNA operations based on the existing answer. It can also quickly solve problems similar to those already solved.

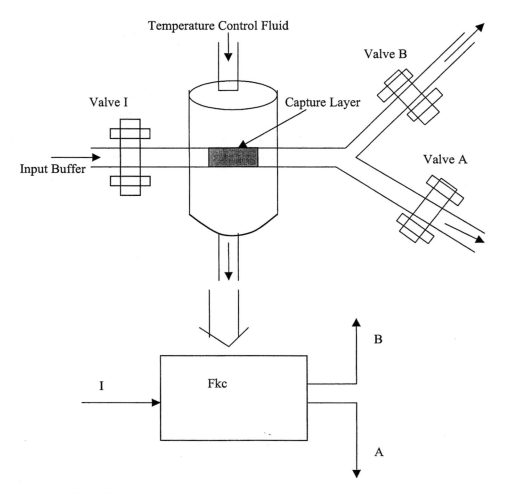

Fig. 10: An example of a filter for the *k*th node with color *C*

Finally, based on the new model, our new algorithms are also presented for decoding all answers to the problem represented by DNA strands. This is a significant advantage over those methods that can locate only a few answers within the whole set. The process of the newly introduced algorithm is very fast and efficient comparing with the existing method using electron microscopy. Instead of only providing the "yes" or "no" answer, the new model can provide exact answers for the problem. Based on the *separate* operation, the new method can decode all the strands in a set with little extra cost. These new algorithms represent a huge improvement over naive search used in the existing algorithms.

No extra material is needed to prepare for the dynamically updating and decoding processes. The only expense is some extra containers for storing the additional pools of DNA strands. As compared to the existing DNA computing algorithms, this new method can achieve a solution much more quickly after the answer for the first problem is generated and it is very financially efficient. This will make DNA computing more attractive to potential users who want to solve problems currently unsolvable.

REFERENCE

[1]   Len Adleman. Molecular computation of solutions to combinatorial problems. *Science*, page 1021-1024, November 1994.

[2]   Len Adleman. *On constructing a molecular computer*. Manuscript, 1995.

[3]   Eric Bach and Anne Condon. DNA models and algorithms for NP-complete problems. *Journal of Computer and System Sciences*, 57:172-186, 1996.

[4]   Ravinderjit S. Braich, Cliff Johnson, Paul W. K. Rothemund, Darryl Hwang, Nocholas Chelyapov, and Leonard Adleman. Solution of a satisfiability problem on a gel-based DNA computer. In *Sixth International Meeting on DNA Based Computers*, pages 31-42, June 2000.

[5]   Nicos Christofides. *Graph Theory: An Algorithmic Approach*. Academic Press, 1975.

[6]   John Clark and Derek Allan Holton. *A First Look at Graph Theory*. World Scientific, 1991.

[7]   Thomas H. Cormen, Charles E. Leiserson, and Ronald L. Rivest. *Introduction to Algorithms*. MIT Press, 1990.

[8]   Y. Gao, M. Garzon, R.C. Murphy, J.A. Rose, R. Deaton, D.R. Franceschetti, and S.E. Stevens Jr. DNA implementation of nondeterminism. In *3rd DIMACS Workshop on DNA Based Computers*, pages 204-211, June 1997.

[9]   Gre Gloor, Lila Kari, Michelle Gaasenbeek, and Sheng Yu. Towards a DNA solution to the shortest common superstring problem. In *Fourth International Meeting on DNA Based Computers*, pages 111-116, June 1998.

[10]   Vineet Gupta, Srinivasan Parthasarathy, and Mohammed J. Zaki. Arithmetic and logic operation with DNA. In *3rd DIMACS Workshop on DNA Based Computers*, page 212-222, June 1997.

[11]   Peter Kaplan, David Thaler, and Albert Libchaber. Parallel overlap assembly of paths through a directed graph. In *3rd DIMACS Workshop on DNA Based Computers*, page 127-141, June 1997.

[12]   Peter D. Kaplan, G. Cecchi, and A. Libchaber. DNA-based molecular computation: template-template interactions in PCR. In *Second Annual Meeting on DNA Based Computers*, pages 159-171, June 1996.

[13]   Richard Lipton. *Using DNA to solve SAT*. Unpublished Draft, 1995.

[14]   Qinghua Liu, Zhen Guo, Anne E. Condon, Robert M. Corn, Max G. Lagally, and Lloyd M. Smith. A surface-based approach to DNA computation. In *Second Annual Meeting on DNA Based Computers*, pages 206-216, June 1996.

[15]   Z. Frank Qiu and Mi Lu. Arithmetic and logic operations for DNA computers. In *Parallel and Distributed Computing and Networks (PDCN'98)*, pages 481-486. IASTED, December 1998.

[16]   Sam Roweis, Erik Winfree, Richard Burgoyne, Nockolas Chelyapov, Myron Goodman, Paul Rothemund, and Leonard Adleman. A sticker based architecture for DNA computation. In *Second Annual Meeting on DNA Based Computers*, pages 1-27, June 1996.

[17] Sam Roweis, Erik Winfree, Richard Burgoyne, Nockolas Chelyapov, Myron Goodman, Paul Rothemund, and Leonard Adleman. A sticker based architecture for DNA computation. In *Journal of Computational Biology*, 1998.

[18] Liman Wang, Qinghua Liu, Anthony Frutos, Susan Gillmor, Andrew Thiel, Todd Strother, Anne Condon, Robert Corn, Max Lagally, and Lloyd Smith. Surface-based DNA computing operations: Destroy and readout. In *Fourth International Meeting on DNA Based Computers*, pages 247-248, June 1998.

[19] Erik Winfree. Proposed techniques. In *Fourth International Meeting on DNA Based Computers*, pages 175-188, June 1998.

[20] Erik Winfree, Xiaoping Yang and Nadrian C. Seeman. Universal computation via self-assembly of DNA: Some theory and experiments. In *Second Annual Meeting on DNA Based Computers*, pages 172-190, June 1996.

In: Biocomputing
Editor: Phillip A. Laplante, pp. 91-112

ISBN 1-59033-889-8
2003 © Nova Science Publishers, Inc.

Chapter 7

A ONE INSTRUCTION SET ARCHITECTURE FOR GENETIC ALGORITHMS

William Gilreath
6224 North Park Meadow Way, Boise, ID 83713
will@williamgilreath.com

Phillip Laplante
Engineering Division
Penn State Great Valley School of Graduate Professional Studies
30 Swedesford Road, Malvern, PA 19355-1443
plaplante@psu.edu

ABSTRACT

Genetic algorithms (GA) are a well-known non-deterministic means to search a problem/solution space to find an optimal but not necessarily best solution within a reasonable time period for computationally intractable problems. The genetic algorithm uses the metaphor of natural evolution with the paramount Darwinian principle of "survival of the fittest."

The genetic algorithm mirrors the evolutionary mechanism in the operations used, namely selection, mutation, crossover and inversion. The genetic algorithm is used to evolve "chromosomes" that represent possible solutions to the problem, and with each succeeding generation of chromosomes in a population, a fitter chromosome ensues.

One instruction set computing is a form of minimalist computing, in which the functionality of a processor is reduced to one instruction, making possible several key improvements in processor design. A specific instruction set is then formed by the orthogonality of parameters with the one instruction. This approach leads to enhancements in processor organization and structure that include scalability of construction, efficiency of design, and simplicity of programming.

The genetic algorithm can be implemented using the one instruction computer architecture. This application of one instruction computing provides a unique and natural combination as a general-purpose means to optimizing problems. Moreover, the simplified construction using one instruction elements, lends itself well to computing in alternate materials such as organic, optical, chemical, or quantum components and nano-materials. Finally, the one instruction methodology has a biologic parallel in the creation of a one instruction computer using living cells.

Key-words: artificial intelligence, evolutionary computing, genetic algorithm, OISC, one instruction computing

1. INTRODUCTION

The One Instruction Set Computer (OISC, pronounced, "whisk") is the penultimate Reduced Instruction Set Computer (RISC)[1]. In OISC, the instruction set consists of one instruction, and then by the orthogonality of the instruction along with composition, a complete set of operations is synthesized as needed, or for the problem domain. This approach is completely contrary to a complex instruction set computer (CISC), which incorporates complex instructions as micro programs within the processor [1].

A processor implemented with one instruction may appear to be lacking the necessary functionality to be seriously considered. Yet, there are interesting advantages in favor of this approach and practical implementations of OISC have been built for many years [12]. Moreover, a OISC is well suited for reconfigurable, or "programmable" architectures since part or all of OISC architecture can be remapped to organize the hardware resources more efficiently for the type of problem being solved. For example, Moore and Morgan built a OISC on a Xilinx field-programmable gate array (FPGA) that can be reconfigured to regenerate missing or faulty modules [2]. A more specific instruction set would limit how programmable a reconfigurable architecture could be, in essence restricting it.

OISC architectures provide an excellent paradigm for implementing traditional von Neumann computers using non-traditional materials. This aspect is important since von Neumann architectures have a much larger body of software engineering results, including programming languages, tools, and methodologies, than non-von Neumann architectures. Simply put, by massive scaling of simple, single instruction elements, a practical computer can be built. This facet has important implications in nanocomputing, optical computing, biological computing, and in other types of computing involving non-traditional materials.

The question arises "why use OISC for genetic algorithms?" First, the OISC paradigm is well suited for GA because the inherently parallel and uniform architecture maps easily into a GA. Moreover, OISC architectures can scale down to integration levels that are much greater than a general purpose, CISC, or even RISC architecture. The simplicity of the OISC machine also increases the possibility of implementation in alternative materials, which is clearly desirable for emulating organic processing.

There is relatively little work published on applying alternative architectures for implementing genetic algorithms. Exceptions include Inoue, et al who designed a single-instruction-multiple-data (SIMD) computer for GA [3], Twardowski, who used an associative architecture for GA [4], and Martin, who used field-programmable gate arrays (FPGAs) for

---

[1] In the literature the acronym "OISC" is used rather to represent One Instruction Set Computer or One Instruction Set Computing rather than "SISC" for "Single Instruction Set Computer" or "SIC" for Single Instruction Computer to avoid confusion with the more common meanings of those acronyms.

implementing genetic algorithms [5]. All of these examples are noteworthy in that the architecture was adapted to optimize the performance of the algorithm.

In this work, however, Inoue et al's architecture is of particular interest because it utilized specific-purpose processors for the selection, crossover, and mutation processors, in much the same way we shall show here. Twardowski's architecture is also relevant because he used field-programmable gate arrays, which are widely gaining use in special-purpose applications.

FPGAs are desirable for implementing any OISC solution because they are inexpensive and can be reconfigured to adapt to the base instruction. Moreover, there exists a rich toolset to develop applications for FPGAs including high order language compilers, simulators and logic minimization software. One would have to develop all of these from scratch if a custom OISC were built.

The approach presented here is, in a sense, a fusion of OISC and FPGA technology because it uses special, single-purpose processors in a one instruction set paradigm, but relies on the availability and ease-of-implementation of FPGA, and accompanying tools for the realization of the computer.

2. GENETIC ALGORITHMS

The genetic algorithm was introduced in the 1970s [6] as a metaphor of evolution through natural selection to find solutions to computationally difficult problems. Genetic algorithms are often used in lieu of a deterministic algorithm, or exhaustive search of the solution space where the time to find a solution would be exponential.

Genetic algorithms use a fitness or objective function to rate the fitness of the primary element of a genetic algorithm, the "chromosome," which represents a possible solution to the problem. A simple chromosome is a binary string, encoding the problem solution; likewise, a fitness function, which is customized to the problem to be solved, ranks the fitness of a chromosome.

There are four operators or operations in the genetic algorithm:

1. Selection
2. Crossover
3. Mutation
4. Inversion

Selection is the means of choosing two chromosomes as parents to generate offspring. The most popular means of selection is the "roulette wheel" approach, where the fitness of all chromosomes in a population is used to determine how likely that chromosome is to be selected [7]. Better-fit chromosomes are more likely to be selected from the existing population. If the same chromosome is selected, then it is simply a pass-through to the next generation of chromosomes. Other techniques include normalizing the chromosomes, or ranking, and tournament selection [8].

Once two chromosomes have been selected they have to be joined to generate two offspring. The crossover operation joins two chromosomes so that offspring are generated. Crossover can be single point, two points, or many points of interchange of information among the chromosomes. At each point along the chromosome, the information is exchanged between each chromosome, creating a new chromosome. The newly generated offspring are then passed into the next generation population. There are other variations on crossover, such as meeting a minimum fitness to be passed into the next generation, or simply copying a chromosome into the next generation.

Often times, a genetic algorithm can prematurely converge to a solution when a chromosome's fitness is too high, and the succeeding generations are copies of the original "super-chromosome" that was generated. The mutation operator causes a random change in a chromosome, to help alleviate the problem of premature convergence to a chromosome. The mutation operation also helps keep genetic diversity, so that a population does not become homogenous. A more diverse population of chromosomes is a more diverse space of candidate solutions to the problem being solved [9].

Holland had proposed the operation of inversion -- randomly shuffling of genes within a chromosome. Inversion has not been used in most genetic algorithms, however, because the operators of crossover and mutation effectively achieve the same effect [6].

The "algorithm" part of a genetic algorithm is the process of applying the genetic algorithm operations consistently, and generating a population of solutions in an iterative way. An initial population of one to two hundred chromosomes is either encoded with possible solutions to the problem being solved, or selected completely at random, or a mix of seeding and random initialization. The genetic algorithm then continues to create a next generation population and test each generation's solution, until a sufficiently optimal solution or the optimal solution is generated.

3. ONE INSTRUCTION SET COMPUTING PROCESSORS

There are several historical paradigms for OISC, which include the subtract-and-branch-if-negative, MOVE, and the half adder. All of these architectures have been shown to be functionally equivalent [10], and it will be asserted that any of these can be used to implement a genetic algorithm. The following sections provide a brief description of these architectures.

3.1. Half adder architecture

The half adder (HA) is a well-known, logical device that can perform addition of two one-bit binary numbers and produce their sum and carry. The sum operation can be easily implemented using an exclusive or (XOR) operation and the carry via a logical AND operation.

It has been shown that a very simple OISC can be built using the half adder as the one instruction and that by employing HAs, any device requiring the AND, OR, XOR and NOT operations can be realized. The program counter, general and special purpose registers and memory locations are realized as part of the address space -- not unlike some commercial processors. Input and output is achieved through memory-mapped I/O or DMA [11].

Although the components of a HA based OISC can be organized in a von Neumann architecture, a parallel organization is more desirable. In particular because of the inherently parallel nature of many image-processing algorithms, a parallel image-processing computer can be built using only the HA element and the logic components derived from it. However, parallelism needs to be planned in the design, and is entirely dependent on the nature of the problem being solved, as will be demonstrated later.

The HA instruction set is complete and can be used to construct any Turing computable program [11]. This means that any practical program can be built using this instruction.

3.1.1 A Cell Based Half Adder Architecture

The possibility of building computers based on simple instructions such as the half adder suggests implementations of such computers in alternate materials or even living organisms. Indeed, others have already suggested that living cells might be used to build a practical computer [12].

One approach to a living computer is based on the assertion that the interaction between a cell, certain proteins, and DNA can be modeled by the partial finite state automaton (FSA) shown in Fig. 1 [13], and the fact that such an automaton can be used to build half adders, which are known to be a workable paradigm for one instruction computers.

The FSA shown in Fig. 1 can be generalized by the FSA shown in Fig. 2.

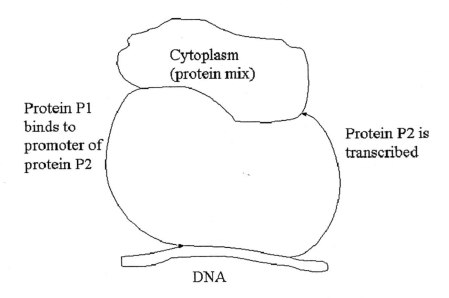

Fig. 1: A finite state automaton that models the interaction between a cell, certain proteins and DNA [13].

Fig. 2 has matrix representation given by

	A	B
α	B	A
β	A	B

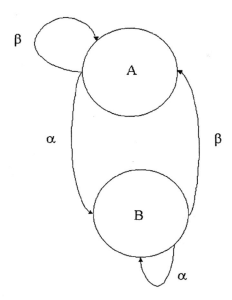

Fig. 2: A generalized version of the cell FSA shown in Fig. 1.

Encoding α=1, β=0, A=0 and B=1, yields

	0	1
1	1	0
0	0	1

Finally, rewriting the column, row and internal entries as a truth table yields

Input	State	Next State
0	0	0
0	1	1
1	0	1
1	1	0

Which is simply the truth table for A XOR B. Hence a cell of this type can behalf as an XOR gate.

Now if the behavior of the FSA is subject to changes in the proteins, as asserted [13], then an alternate representation for Fig. 1 is shown in Fig. 3.

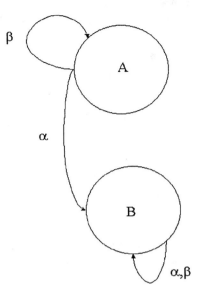

**Fig. 3: An alternate representation for the cell/
protein/DNA interaction shown in Fig. 1.**

Now the matrix representation for the FSA in Fig. Z is

	A	B
α	B	B
β	A	B.

Encoding α=0, β=1, A=1 and B=0, yields

	1	0
0	0	0
1	1	0

Finally, rewriting the column, row and internal entries as a truth table yields

Input	State	Next State
0	0	0
0	0	0
1	0	0
1	1	1

Which is simply the truth table for A AND B. Hence a cell of this type appears to behave like an AND gate.

This leads to the conclusion that, depending on the cell/protein/DNA interactions asserted, it appears to be possible to implement a half adder using the AND and XOR cell behaviors.

3.2 Subtract-and-branch-if-negative instruction OISC processor

The subtract-and-branch-if-negative (SBN) processor was originally proposed by van der Poel [4] and has been repeated throughout the literature, including in mainstream texts on computer architecture [15]. The instruction is of the format:

SBN operandam, operandum, next-address

Here, the operandum is subtracted from the operandam, and if the result is negative, execution proceeds to the instruction stored at the address in the branch field. Otherwise, execution proceeds with the next instruction immediately stored in memory. That is, the program counter increments. Written as pseudo-code the SBN instruction takes the form:

operandam = operandam – operandum;

if(operandam < 0)
goto next-address;
fi;

The SBN instruction (Singleton) set is complete and can be used to construct any Turing computable program [1].

3.3. MOVE Based Processor

A third, and indeed most practical, paradigm for OISC computing involves the MOVE instruction. Most practical OISC implementations have been constructed using variations of this architecture. MOVE OISCs rely on a single data transfer operation, "MOVE" and achieve diverse functionality by memory–mapped operations implemented at the target memory locations. In a basic MOVE architecture, the computer consists of several ALU, conditional unit, I/O Devices, and memory-mapped registers. The MOVE operation to the appropriate unit triggers an arithmetic, load or store operation (see Fig. 4). Such an approach is also known as a "transport-triggered" architecture [16]. While a MOVE architecture is really a multi-instruction architecture, it appears as a single instruction to compiler and programmer.

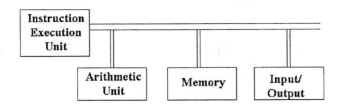

Fig. 4: A MOVE Architecture [16].

The MOVE instruction is of the format of:

MOVE operandam, operandum

In such an instruction, the contents of the operandam are moved to the operandum, but the actual computation is achieved by underlying hardware abstracted as memory at the operandum location. In essence, the memory location address only identifies functionality, so the actual functionality is memory-mapped.

Written as pseudo-code, the MOVE instruction takes the form:

operandum := operandam;

or more functionally:

copy(operandum, operandam);

The MOVE instruction set is complete and can be used to construct any Turing computable program [1]. However, different digital structures are need to be developed, or existing ones used, for each instruction type to be implemented. For example, in Fig. 4 the Instruction Execution Unit is actually an abstraction for ADD, MULT, SUM, and other CLBs interconnected in a continuous organization.

4. IMPLEMENTING A GENETIC ALGORITHM PROCESSOR/ENGINE

The OISC computer architecture with its generality and simplicity makes implementing a genetic algorithm processor or engine not only possible, but also architecturally appropriate. In effect, the functionality of the genetic algorithm operators becomes memory addresses -- the genetic algorithm operators implemented in hardware. The organization of the memory addresses creates virtual registers, which are used to implement the algorithm. The genetic algorithm itself is based on a series of MOVE instructions that load and store the results. Other instructions can be synthesized to perform comparisons, loops, branches, and other instructions [1]. Fig. 5 illustrates the architecture for a MOVE-based Genetic Algorithm processor.

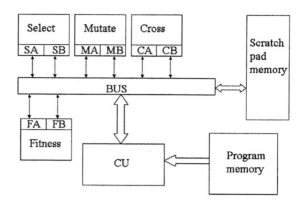

Fig. 5: An architecture for a MOVE-based OISC.

The crux of the processor is the implementation of the genetic algorithm operators. The remaining instructions can be synthesized to provide ancillary support.

The following sections describe much of the detailed implementation of the genetic algorithm using MOVE-based OISC. While a simulator has been built and tested (see Appendix A) we discuss only the necessary C-code and MOVE-based OISC equivalent to illustrate the feasibility of the approach. The MOVE-based OISC code is indented and preceded by the mnemonic "asm" for readability. In addition, most of the prologue, epilogue and variable declarations are omitted for brevity.

4.1. Chromosome Design

Each chromosome is a binary word in memory where each binary substring is assigned to one of the variables in the problem to be optimized. For instance, 128 bits will be used in the running example to represent two 64 bit variables. When the algorithm completes, the chromosome can be decoded back to the problem parameters.

The C declaration for the chromosome is given below and is provided to increase the readability of subsequent C code and its MOVE-based OISC equivalent.

```
/* sizeof chromosome is 2 64-bit words or 16-bytes */
typedef struct
{
float fitness; /* 64-bit float */
long gene; /* 64-bit long */
} chromosome; /* end chromosome */
```

Naturally, in a real OISC computer the allocation of the data structures needs to be taken care of manually. But in the case of using FPGAs, the translation software takes care of this (and is described later).

The chromosome population is initialized using a simple looping construct and random number generator.

4.2. Objective or Fitness Function

The objective or fitness function is represented as two memory addresses, one for the chromosome (address FA in Fig. 5), and another for the fitness value (address FB). After moving a chromosome into the fitness register, the fitness value is generated and placed in the fitness value register. In a 64-bit word system, the fitness value can be a floating-point value between 0.0 and 1.0.

The equivalent C code and MOVE-based OISC for the fitness function is:

```
#define MUTATION_RATE 0.01
#define PASSTHRU_RATE 0.05

void eval_fitness(chromosome *c)
{
asm Move c->gene, FITNESS_REG;
asm Move FITNESS_REG, c->fitness;

}/* end fitness */
```

4.3. Selection

Selection is implemented with two memory addresses that act as registers. The first(address SA in Fig. 5) is the selection base memory address for the chromosome population, where chromosomes reside. The second address (address SB) is the population size register, which indicates how many chromosomes are in the population.

Selection is performed iteratively by a loop, which starts at a random location between the base address and the end address, by computing a random offset between 0 and the population size. The following C code followed by its MOVE equivalent illustrates selection.

```
chromosome selection(chromosome c[])
{
chromosome select;
select.gene = 0L;
select.fitness = 0.0f;

asm Move &c, SELECT_POP_FIRST_REG;
asm Move POP_SIZE + &c, SELECT_POP_LAST_REG;
asm Move SELECT_REG, select->gene;

eval_fitness(&select);
```

return select;

}/* end selection */

Another random value is the minimum accumulated fitness, $0.0 \le r \le 1.0$. Once the random starting point and the minimum accumulated fitness are computed, a chromosome is moved into the fitness register. Then the fitness value is added and accumulated.

If the fitness value accumulated is less than the minimum fitness, the selection process continues to the next memory address. The next chromosome is then selected in the same manner. Once two chromosomes have been selected, crossover then occurs. After selecting one chromosome, there is a probability that the one chromosome will be chosen for mutation, in which case it is moved into the mutation register, then into the next generation population.

Another register indicates the base address of where the base population starts, and another register indicates a count of the number of chromosomes in the next generation population, which is initially set to zero.

4.4. Crossover

Crossover uses two registers, one for each chromosome. When two chromosomes are moved into the registers, the resultant chromosomes are stored in two output registers. Otherwise, each cycle would cause crossover for the offspring if they were put in the original registers. Crossover is commutative, that is, chromosome X and chromosome Y can be transposed, with the same offspring as a result. Crossover is illustrated in the following C code and MOVE OISC equivalent:

```
chromosome selection(chromosome c[])
{
chromosome select;
select.gene = 0L;
select.fitness = 0.0f;

asm Move &c, SELECT_POP_FIRST_REG;
asm Move POP_SIZE + &c, SELECT_POP_LAST_REG;
asm Move SELECT_REG, select->gene;

eval_fitness(&select);

return select;
}/* end crossover */
```

The crossover mechanism used is ideally uniform. The crossover register will randomly interchange each bit in the parent chromosomes binary bits, and then move the offspring into the next generation population. This is illustrated by the following code:

```
chromosome *crossover(chromosome *x, chromosome *y)
```

```
{
chromosome *offspring = malloc(CHROM_SIZE);
offspring->gene = 0L; /* offspring->gene= 0L; */
offspring->fitness = 0.0;
asm Move x->gene, CROSSOVER_OPA_REG; /* Move 64-bit long int crossover
operandam */
asm Move y->gene, CROSSOVER_OPU_REG; /* Move 64-bit long into crossover
operandum */
asm Move OFFSPRING_REG, offspring->gene;
eval_fitness(offspring);
return offspring;
}/* end crossover */
```

4.5. Mutation

If a chromosome is chosen to be mutated, the chromosome is loaded into the mutation register, which randomly inverts bits of the chromosome. Random bit selection has a 0.5 probability of mutating a bit to its current value, in effect, decreasing the probability for mutation on a bit by one half. Once mutation is complete, the chromosome is moved into the next generation population.

As each chromosome is moved into the next generation population, the fitness can be computed, and compared to a desired fitness parameter in a memory location. When the algorithm generates a chromosome that meets the minimum fitness, the algorithm can stop, or the chromosome could be moved into the next generation population. The same chromosome could then be sent to an output register, which contains the best possible solution to the problem so far. Mutation can be coded in the following way:

```
chromosome *mutate(chromosome* c)
{
asm Move c->gene, MUTATE_REG; /* move the gene 64-bit long into the mutation
register */
asm Move MUTATE_REG, c->gene; /* move the mutated gene back into the
chromosome */
eval_fitness(c);
return c;
}/* end mutate */
```

4.6. Universal Implementation

It should be restated that the Select, Crossover and Mutation processors can themselves be implemented using MOVE or SBN based OISC. Similarly, the GA OISC architecture could be implemented using SBN elements or both MOVE and SBN, since it has been shown that either element can be implemented using the other [1].

5. FIELD PROGRAMMABLE GATE ARRAYS

In looking for a practical platform for implementing Single Instruction Computing, and recognizing that it is an "exotic" architecture and not ready for mass production, one is naturally drawn to the FPGA. Field programmable gate arrays have reached usable densities in the hundreds of thousands of gates. They provide gates and flip-flops, which can be integrated to form a system level solution. Clock structures can be driven using dedicated clocks that are provided within the system. FPGAs are infinitely reprogrammable (even within the system) and design modifications can be made quickly and easily [17]. Hence, they are well adapted to the massed digital logic approach of OISC.

Programmable logic consists of a series of logic blocks connected in either segmented or continuous interconnect structures (see Fig. 6). Segmented interconnections are used for short signals between adjacent configurable logic blocks (CLB), while continuous interconnections are used for bus-structured architectures [17].

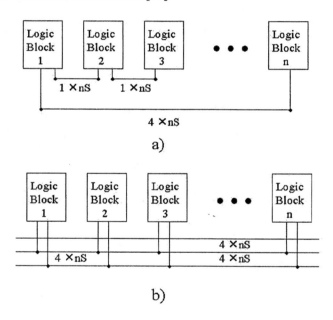

**Fig. 6: Segmented (a) and continuous (b) interconnection
strategies for FPGA logic blocks [18].**

The continuous structure is ideal for connecting large numbers of simple logic units, such as half adders, full-adders and twos complement gates. Each of these can easily be constructed from a single logical operation [17]. Moreover, these logic blocks can also be predesigned to implement higher level functions, such as vector addition, convolution, or even FFT. Indeed, the SBN instruction can be directly implemented as a CLB by cascading existing CLBs for subtraction and branching [18]. The ability to reconfigure logic blocks gives the flexibility to select a single instruction and use it to implement the necessary functions. This is a key strength of an FPGA solution to implementing a OISC for experimental and practical implementation purposes.

Implementation of memory decoders, DMA, clocks, etc. using FPGAs is straightforward because manufacturers provide these as core logic elements. Since all the basic logic functions have been provided, realization of any standard computer hardware is just an exercise in patience and large-scale integration. Since each HA is a simple CLB interconnected in continuous fashion, the HA architecture is perhaps the closest to existing logic implementations using FPGAs.

6. DEVELOPMENT ENVIRONMENT

A rich development environment is available for producing CLBs for standard commercial FPGAs. Fig. 7 (a) illustrates the traditional development cycle of Verilog/VHDL programming to produce FPGA Net lists. VHDL is the acronym for VHSIC (Very High Speed Integrated Circuit) Hardware Description Language.

Verilog® is a popular language used in synthesizing designs for FPGAs. While it is less verbose than traditional VHDL and is now being standardized by the IEEE 1364 working group, it was not originally intended as an input to synthesis since many Verilog constructs are not supported by synthesis software [18].

VHSIC Hardware Description Language (VHDL) is a hardware description language used for designing integrated circuits. Like Verilog, it was not intended for synthesis. However, the high level abstraction of VHDL makes it easier to describe system components that are not synthesized [15].

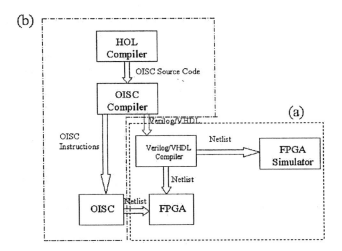

Fig. 7: Mapping OISC into VHDL then to an FPGA. (a) represents the existing software support. (b) represents new support tools needed for OISC.

Another advantage of using standard languages like VHDL and Verilog is that gate reduction can be realized through the use of a commercial solution, such as Synopsys® DesignWare.

Fig. 7 (b) illustrates the software support tools needed for a OISC system. The HOL compiler takes source code, typically in a language like C and converts it to OISC symbolic code. The OISC compiler converts the symbolic code either directly into native instructions for the machine, or into Verilog or VHDL. The purpose of the alternate compilation path is to permit simulation of the code on FPGA simulators, or for direct mapping to an FPGA.

Because of standardization, implementations are portable across commercial FPGA types and this further expands the set of available CLBs. Table 1 shows the applicable standards for VHDL, Verilog and for the Std_logic data type used in programming the FPGA using version of standard ANSI C.

Table 1: Industry standards for FPGA programming [18].

Standard	Version
VHDL Language	IEEE-STD-1076-87
Verilog Language	IEEE-STD-1364-95
Std_logic Data Type	IEEE-STD-1164-93

7. SUMMARY AND CONCLUSIONS

In this paper, a one instruction or OISC computer was used to implement the genetic algorithm. While the focus was on the MOVE-based OISC, the SBN instruction and half adder architecture were also investigated. In particular, a theoretical implementation of the half adder architecture using cells was described. The half adder can be used to implement SBN or MOVE-based OISC, making this cellular architecture potentially practical.

Finally, the use of FPGA devices in implementing prototype systems was examined as mechanism for proof-of-concept. Using an FPGA, the functionality of the OISC processor can be implemented by describing it in a C-like programming language, such as VHDL or Verilog, then tested before trying to construct such a system in alternate materials.

REFERENCES

[1]   William F. Gilreath and Phillip A. Laplante, *Computer Architecture: A Minimalist Approach*, to be published by Kluwer Academic Press, December 2002.

[2]   S. W. Moore and G. Morgan "The Recursive MOVE Machine: r-move", *IEE Colloquium on RISC Architectures and Applications*, 1991, pp. 3/1 – 3/5.

[3]   T. Inoue, M. Sano and Y. Takahashi, "Design of a Processing Element of a SIMD Computer for Genetic Algorithms," High Performance and Computing on the Information Superhighway, HPC, 1997, pp. 688-691.

[4]   Kirk Twardowski, "An Associative Architecture for Genetic Algorithm-Based Machine Learning," *Computer*, Volume 27, Issue 11, November 1994, pp. 27-38.

[5]   Peter Martin, "A Hardware Implementation of a Genetic Programming System Using FPGAs and Handel-C," *Genetic Programming and Evolvable Machines*, Vol. 2, 2001, pp. 317-343.

[6]   J. Holland, *Adaptation in Natural and Artificial and Natural Systems*, University of Michigan Press, Ann Arbor, 1975.

[7]   J. E. Baker, "Adaptive selection methods for genetic algorithms," *Proceedings of the International Conference Genetic Algorithms and Their Applications*, Pittsburgh, PA. 1985, pp. 101-111.

[8]   Tobias Blickle and Lothar Thiele. A comparison of selection schemes used in genetic algorithms. Technical Report 11, Computer Engineering and Communication Networks Lab (TIK), Swiss Federal Institute of Technology (ETH) Zurich, Gloriastrasse 35, CH-8092 Zurich, 1995.

[9]   Marcus Hutter. Fitness Uniform Selection to Preserve Genetic Diversity. Proceedings of the 2002 Congress on Evolutionary Computation (CEC-2002) 783--788, *IEEE Transactions Evolutionary Computation*. 2002.

[10] William F. Gilreath and Phillip A. Laplante, "A Basis for Single Instruction Set Computing", in review, *Journal of Systems Architecture*, March 2002.

[11] P. A. Laplante, "A Novel Single Instruction Computer Architecture, " *ACM Computer Architecture News*, vol. 18, no. 4, December 1990, pp. 22-23.

[12] Katie Pennicott, "'DNA' Computer Cracks Code," *Physics Web*, http://physicsweb.org/article/news/6/3/11, published March 15, 2002, accessed 3/18/2002.

[13] Dror G. Feitelson and Millet Treinin, "The Blueprint for Life?," IEEE Computer, July 2002, pp. 34-39.

[14] W.L. van der Poel, "The Essential Types of Operations in an Automatic Computer," *Nachrichtentechnische Fachberichte*, vol. 4, pp. 144-145, 1956.

[15] John Hennessy and David Patterson, *Computer Organization and Design: The Hardware/Software Interface.* 2nd edition, Morgan-Kaufmann Publishers, San Francisco, CA. 1998.

[16] Douglas Jones, "The Ultimate RISC," *Computer Architecture News*, 1985, pp. 48-55.

[17] *Synopsys (XSI) Synthesis and Simulation Design Guide*, Xilinx, Inc., San Jose, CA, 1997.

[18] *Core Solutions Databook*, Xilinx, Inc., San Jose, CA, 1998.

APPENDIX A C CODE AND MOVE OISC EQUIVALENT

```
/****
Authors: William Gilreath && Phillip Laplante
Date: September 2002
Source file: oisc_sga.c
```

C code for simple genetic algorithm that runs on a theoretical 64-bit machine one-instruction computer using Move with specific register functionality to implement the genetic algorithm operations of:

selection : SELECT_POP_FIRST_REG, SELECT_POP_LAST_REG

crossover : CROSSOVER_OPA, CROSSOVER_OPU, OFFSPRING_REG

mutation : MUTATION_REG
fitness : FITNESS_REG

```
fitness : FITNESS_REG

****/

#include <stdio.h>
#include <stdlib.h>

/* sizeof chromosome is 2 64-bit words or 16-bytes */
typedef struct
{
float fitness; /* 64-bit float */
long gene; /* 64-bit long */

} chromosome; /* end chromosome */

/* population size and chromosome bytesize */

#define POP_SIZE 100
#define CHROM_SIZE sizeof(chromosome)

/* simple genetic algorithm paramters for pass through and mutation rate */

#define MUTATION_RATE 0.01
#define PASSTHRU_RATE 0.05

#define asm _asm NOP ; //macro to nullify asm to compile

void eval_fitness(chromosome *c)
{
asm Move c->gene, FITNESS_REG;
asm Move FITNESS_REG, c->fitness;


}/* end fitness */


void init(chromosome pop[])
{
int x = -1;
srand(123456789L); /* random seed for long generator */

/* initialize all members of the population */
for(x=0; x < POP_SIZE; x++)
{
 pop[x].gene = rand(); /* initial random long */
 eval_fitness(&pop[x]); /* compute fitness */
```

```
}/* end for */
}/* end init */

chromosome *mutate(chromosome* c)
{
asm Move c->gene, MUTATE_REG; /* move the gene 64-bit long into the mutation
register */
asm Move MUTATE_REG, c->gene; /* move the mutated gene back into the
chromsome */

eval_fitness(c);

return c;

}/* end mutate */

chromosome *crossover(chromosome *x, chromosome *y)
{
chromosome *offspring = malloc(CHROM_SIZE); /* chromosome offspring* =
malloc(CHROM_SIZE * 1); */
offspring->gene = 0L; /* offspring->gene = 0L; */
offspring->fitness = 0.0; /* offspring->fitness = 0f; */

asm Move x->gene, CROSSOVER_OPA_REG; /* Move 64-bit long into crossover
operandam */
asm Move y->gene, CROSSOVER_OPU_REG; /* Move 64-bit long into crossover
operandum */
asm Move OFFSPRING_REG, offspring->gene;

eval_fitness(offspring);

return offspring;

}/* end chromosome */

chromosome selection(chromosome c[])
{
chromosome select;
select.gene = 0L;
select.fitness = 0.0f;

asm Move &c, SELECT_POP_FIRST_REG;
asm Move POP_SIZE + &c, SELECT_POP_LAST_REG;
asm Move SELECT_REG, select->gene;

eval_fitness(&select);
```

```
return select;


}/* end selection */

/* find the maximum or best chromosome in the population */ chromosome
maximum(chromosome pop[]) {
    int x = -1;
    chromosome max;

    max.gene = 0L;
    max.fitness = 0.0;

    for(x=0; x<POP_SIZE; x++)
    {
    max = pop[x];

    if(max.fitness < pop[x].fitness)
    {
    max = pop[x];

    }/* end if */

    }/* end for */

    return max;

}/* end maximum */


/**

main routine, called as: oisc_sga [0.0 < <target fitness> < 1.0]

**/

void main(int argc, char* argv[])
{
chromosome *population = malloc(POP_SIZE * CHROM_SIZE);
chromosome *next_gen = malloc(POP_SIZE * CHROM_SIZE);


int x = -1;

float chance = -1.0; /* random determination */
float targetFitness = -1.0;
```

```
chromosome maxChromosome, first, last, *result;

init(population); /* initialize population */
srand(987654321); /* seed random generator */

/* initialize maximum chromsome */
maxChromosome.fitness = 0.0;
maxChromosome.gene = 0L;

/* check command args for target fitness */
if(argc <= 1)
{
printf("Usage: sga <target_fitness 0.0 ... 1.0>");
exit(1);
}
else
{
targetFitness = (float)atof(argv[1]); /* get desired target fitness */
}/* end if */


while(maxChromosome.fitness < targetFitness)
{
for(x=0; x < POP_SIZE; x++)
{
first = selection(population); /* select parent from population as index */

chance = 1.0; //(float)rand()/MAX_RAND; /* randomly determine operation */

if(chance <= MUTATION_RATE)
{
                                                result = mutate(&first);
                                                next_gen[x] = *result;

continue;
}
else
if(chance <= PASSTHRU_RATE)
{
                                                result = &first;
 next_gen[x] = *result;
continue;
}
else
{
 last = selection(population);
```

```
                                           result = crossover(&first,&last);
                                               next_gen[x] = *result;
        }/* end if */


        }/* end for */


        }


    printf("Best chromosome has a fitness of %-10f", maxChromosome.fitness); /* printf
left-justified float with minimum field of 10 characters */
    printf("Chromsome bit-pattern in hex is %-10x", maxChromosome.gene); /* printf as
hexadecimal with minimum field of 10 characters */

    exit(0);

    }/* end main */
```

INDEX